Examples in
A-LEVEL CORE MATHEMATICS

Also available from Stanley Thornes

Examples in
A-LEVEL CORE MATHEMATICS

Ewart Smith MSc

Head of Mathematics Department
Tredegar Comprehensive School

Stanley Thornes (Publishers) Ltd

Text © Ewart Smith 1990
Original line illustrations © Stanley Thornes (Publishers) Ltd 1990

First published in 1990 by:
Stanley Thornes (Publishers) Ltd
Old Station Drive
Leckhampton
CHELTENHAM GL53 0DN
England

Reprinted 1992

British Library Cataloguing in Publication Data

Smith, Ewart
    Examples in A-level core mathematics.
    I. Title
    510

    ISBN 0-7487-0440-X

Typeset by Tech-Set, Gateshead, Tyne & Wear.
Printed and bound in Great Britain at The Bath Press, Avon.

Contents

Preface

This book aims to cover the core course in A-level mathematics. Additional work is included so that virtually all the topics currently appearing on the basic pure mathematics papers of the various Examining Boards are supplemented. The book is divided into three parts: Part 1 gathers useful facts together according to topic. Numerous questions of varying degrees of difficulty follow each set of facts or topic title. Part 2 consists of ten revision papers, each paper containing twelve questions on a variety of topics. These questions are of examination standard. Part 3 reproduces questions from recent A-level papers of the various Examining Boards. They span the whole range of topics normally examined and vary in degree of difficulty. Answers are provided at the end of the book.

Many people have helped in the creation of this book. In particular I acknowledge the work done by my colleague Allan Snelgrove in checking my solutions; John Roberts for his comments and encouragement; David Jenkins and the members of his department at Olchfa C.S., Swansea, for reading the manuscript and making many useful suggestions. Above all I would thank my A-level students throughout the last thirty years from whom I have learnt so much.

I am grateful to the following Examining Boards for permission to reproduce questions from past papers:

The Associated Examining Board (AEB)
University of Cambridge Local Examinations Syndicate (C)
Joint Matriculation Board (JMB)
Northern Ireland Schools Examination Council (NI)
Oxford and Cambridge Schools Examination Council (O&C)
Southern Universities' Joint Board (SU)
Welsh Joint Education Committee (W)

All answers to examination questions are provided by myself, they are not the responsibility of the Boards.

I must also thank my publishers for the care and attention to detail they have taken at each stage of production. Finally my thanks are due to my wife and family for encouraging me to spend so much time in my study.

Ewart Smith
1990

Part 1: Useful Facts and Exercises

Manipulative Algebra 1

EXERCISE 1a

Factorise:

1 $2x^2 + 7x + 3$

2 $2x^2 + 5x - 3$

3 $5x^2 - 22x + 8$

4 $4x^2 + 8x + 3$

5 $16x^2 - 16x + 3$

6 $18x^2 - 9x - 20$

7 $16a^2 - 8a - 3$

8 $15x^2 + 4x - 35$

9 $8x^2 - 2x - 15$

10 $9a^2 + 3a - 20$

11 $2x^2 - 2x - 12$

12 $3x^2 + 36x + 105$

13 $9 + 21a + 12a^2$

14 $12x^2 + 20x - 8$

15 $a^2 - 25b^2$

16 $48x^2 - 3$

17 $16 - 100a^2$

18 $12x^2 - 3$

19 $a + b + ab + 1$

20 $2ab + 2a - 2b - 2$

21 $12 - 7x + x^2$

22 $20 - 3x - 2x^2$

23 $21 + 4a - a^2$

24 $15 + 2b - b^2$

25 $2a^2 - a + 2ab - b$

26 $mn - 3m + 2n^2 - 6n$

27 $2pq + 4p + 3q + 6$

28 $2pq + 4p - q - 2$

29 $ax - bx + ay - by$

30 $ax - 2ay + 2by - bx$

31 $a^2 - b^2 + 3(a + b)$

32 $3(a - b) - (a^2 - b^2)$

33 $a^2 + 2ab + b^2 + 2a + 2b$

34 $a^2 - 2a - b^2 - 2b$

35 $a^3 + 1$

36 $8x^3 + 1$

37 $a^4 - a$

38 $x^6 + 8y^6$

39 $3 - 3000x^3$

40 $16a^3 - 54b^3$

EXERCISE 1b

Simplify:

1 $\dfrac{5a^2}{3bc} \times \dfrac{6bc^2}{10ab^2}$

2 $\dfrac{a}{b} \times \dfrac{b}{c} \times \dfrac{c}{a}$

3 $\dfrac{15x^2}{4yz} \times \dfrac{8xy^2z}{3} \times \dfrac{2z^3}{5x^2y}$

4 $\dfrac{5bc^2}{3a} \times \dfrac{3}{25abc} \times \dfrac{a^2}{4c}$

5 $\dfrac{6a^2b}{5c} \times \dfrac{20a^2c}{b} \div 12a^3$

6 $6xyz \times \left(\dfrac{y}{2xz}\right)^2 \div \dfrac{y^4}{2xz}$

7 $\dfrac{a^2 - 9}{a + 3}$

8 $\dfrac{2a^2 + 7a + 6}{3a^2 + 11a + 10}$

9 $\dfrac{8a^2 - 2a - 3}{6a^2 - 5a - 4}$

10 $\dfrac{a^2 + a - 2}{a^2 - a} \times \dfrac{3a - a^2}{a^2 - a - 6}$

11 $\dfrac{x^2 - x - 12}{x^2 + x - 6} \times \dfrac{3x - 6}{x^2 + 2x - 24}$

12 $\dfrac{a^2 - 1}{a - 1} \times \dfrac{a^2 - a - 6}{a^2 + 3a + 2}$

13 $\dfrac{3a + 9}{a^2 - 1} \div \dfrac{a + 3}{a - 1}$

14 $\dfrac{x^2 + 2x - 8}{x^2 + x - 6} \div \dfrac{x^2 + 9x + 20}{x^2 + 3x}$

15 $\dfrac{1}{a + b} + \dfrac{1}{a - b}$

16 $\dfrac{1}{x + y} - \dfrac{1}{x^2 - y^2}$

17 $\dfrac{3}{4 - 3x} + \dfrac{4 + 3x}{3}$

18 $\dfrac{x}{x + 1} - \dfrac{x^2}{x^2 - 1}$

19 $\dfrac{4}{4 - a^2} + \dfrac{1}{a - 2}$

20 $\dfrac{3}{a^2 - 9} - \dfrac{1}{3 - a}$

21 $\dfrac{1}{1 - x} + \dfrac{x}{x - 1}$

22 $\dfrac{4}{3(x + y)} - \dfrac{3}{4(x + y)}$

23 $\dfrac{1}{x + 1} - \dfrac{1}{x - 1}$

24 $\dfrac{1}{x - 1} - \dfrac{2x}{x^2 - 1}$

25 $\dfrac{1}{x - 1} - \dfrac{1}{x + 2}$

26 $\dfrac{2}{x + 1} - \dfrac{2}{x + 2}$

27 $\dfrac{5}{x + 3} - \dfrac{3}{x + 1}$

28 $\dfrac{3}{x - 3} - \dfrac{5}{2x + 1}$

29 $\dfrac{a + 1}{a + 3} - \dfrac{a + 4}{a + 2}$

30 $\dfrac{2a}{a + 3b} + \dfrac{3b}{a - 2b}$

EXERCISE 1c

Solve each of the following equations for the letter shown in brackets.

1 $P = 2a + 2b$ (b)

2 $E = Ri^2$ (i)

3 $2s = a + b + c$ (a)

4 $A = \dfrac{(a + b)}{2}h$ (a)

5 $y^2 = 4ax$ (a)

6 $v = u + at$ (a)

7 $P = L + 2\pi R$ (R)

8 $V = \frac{1}{3}\pi r^2 h$ (h)

3

9 $A = \frac{1}{2}r^2\theta$ (θ)

10 $\dfrac{W}{p} = \dfrac{a}{b}$ (p)

11 $y = mx + c$ (m)

12 $C = 90 - \dfrac{A}{2}$ (A)

13 $v^2 = u^2 + 2as$ (a)

14 $v^2 = u^2 + 2as$ (u)

15 $\dfrac{1}{a} + \dfrac{1}{b} = \dfrac{1}{c}$ (c)

16 $\dfrac{1}{a} + \dfrac{1}{b} = \dfrac{1}{c}$ (b)

17 $a = \sqrt{b^2 - c^2}$ (c)

18 $V = \frac{1}{3}\pi r^2 h$ (r)

19 $mg - T = ma$ (m)

20 $a^2 = b^2 + c^2$ (c)

21 $S = \left(\dfrac{u + v}{2}\right)t$ (v)

22 $A = \pi(R^2 - r^2)$ (r)

23 $A = 2\pi r(r + h)$ (h)

24 $Ax + By + C = 0$ (y)

25 $S = \dfrac{a}{1 - r}$ (r)

26 $E = \frac{1}{2}m(v^2 - u^2)$ (v)

27 $A = P\left(1 + \dfrac{r}{100}\right)$ (r)

28 $C = \dfrac{nE}{R + nr}$ (r)

29 $F = \dfrac{9C}{5} + 32$ (C)

30 $T = 2\pi\sqrt{\dfrac{L}{g}}$ (g)

EXERCISE 1d

Solve each of the following equations for x.

1 $ax + by + c = 0$

2 $y = mx + c$

3 $c^2 = a^2 + b^2 - 2abx$

4 $\dfrac{x}{a} + \dfrac{y}{b} = 1$

5 $y^2 = 4ax$

6 $\dfrac{x - a}{b} = \dfrac{y - c}{d}$

7 $y = \dfrac{x + a}{x + b}$

8 $y = \dfrac{a}{x} - b$

9 $\dfrac{1}{x} = \dfrac{1}{a} + \dfrac{1}{b}$

10 $\dfrac{x - b}{a} + \dfrac{x - a}{b} = 2$

11 $\dfrac{x - a}{a} + \dfrac{x - b}{b} = \dfrac{a}{b} + \dfrac{b}{a}$

12 $ax + b^2 = a^2 - bx$

In questions 13 to 18 solve the given pair of simultaneous equations for x and y.

13 $x + y = 5a$
  $x - y = 3a$

14 $2x + 3y = 3b$
  $x - 3y = 6b$

4

15 $2x + 3y = 9a$
$x - 2y = a$

16 $\dfrac{x}{3} + \dfrac{y}{2} = 4p$

$\dfrac{x}{2} + y = 7p$

17 $x + y = 2a$
$x - y = 2b$

18 $2x + ay = 5ab$
$5x - ay = 2ab$

19 If $Mg - T = Ma$
and $T = ma$

show that $a = \dfrac{M}{M + m} g$

20 If $3v_1 + v_2 = 3u$
and $v_1 - v_2 = -\frac{1}{2}u$
show that $v_1 = \frac{5}{8}u$ and $v_2 = \frac{9}{8}u$

21 If $2v_1 + 3v_2 = 2u_1 + 3u_2$
and $v_1 - v_2 = -\frac{1}{2}(u_1 - u_2)$
find v_1 and v_2 in terms of u_1 and u_2.

22 If $2v_1 + v_2 = 2u$
and $v_1 - v_2 = -eu$
find v_1 and v_2 in terms of e and u.

23 If $y = m_1 x + c_1$
and $y = m_2 x + c_2$

show that $x = \dfrac{c_1 - c_2}{m_2 - m_1}$

and $y = \dfrac{m_2 c_1 - m_1 c_2}{m_2 - m_1}$

What is the value of y when $c_1 = c_2 = c$?

24 If $5g - T = 5a$
and $T - 3g = 3a$
find a in terms of g. Hence find T in terms of g.

25 If $mg - T = ma$
and $T - Mg = Ma$
show that $a = \dfrac{m - M}{m + M} g$ and that $T = \dfrac{2mM}{m + M} g$

26 If $R = \mu_1 S$
and $S + \mu_2 R = W$

show that $S = \dfrac{W}{1 + \mu_1 \mu_2}$

and that $R = \dfrac{\mu_1 W}{1 + \mu_1 \mu_2}$

27 If $S = \frac{1}{3}R$ and $R + \frac{1}{3}S = 2W$
find R and S in terms of W.

28 If $mg \sin \alpha - T = ma$
and $T - Mg \sin \beta = Ma$
find an expression for a independent of T.

29 Given that $P \cos \alpha + F \sin \alpha = Mg \sin \beta$
and $P \sin \alpha - F \cos \alpha = Mg \cos \beta$
show that $P = Mg \sin (\alpha + \beta)$

30 Eliminate u between the equations

$$v^2 = u^2 + 2as \quad \text{and} \quad v = u + at$$

to give a quadratic equation in t. Hence show that $t = \dfrac{v \pm \sqrt{v^2 - 2as}}{a}$

Quadratic Equations   2

If α and β are the roots of the equation $ax^2 + bx + c = 0$

then $$\alpha + \beta = \frac{-b}{a} \text{ and } \alpha\beta = \frac{c}{a}$$

The equation $ax^2 + bx + c = 0$ has real roots if $b^2 \geqslant 4ac$, equal roots if $b^2 = 4ac$, and imaginary roots if $b^2 < 4ac$.

EXERCISE 2

1 If α and β are the roots of the equation $x^2 + px + q = 0$, show that $\alpha + \beta = -p, \alpha\beta = q$.

2 The roots of the equation $x^2 + px + q = 0$ are 2 and 3. Find p and q.

3 The roots of the equation $x^2 + px + q = 0$ are $2 \pm \sqrt{3}$. Find p and q.

4 Form the equation whose roots are 3 and -4.

5 Form the equation whose roots are $5 \pm \sqrt{7}$.

6 Form the equation whose roots are $\frac{1}{2}$ and $-\frac{3}{2}$.

7 Determine whether the roots of the following equations are real or imaginary.
a) $2x^2 + 5x + 2 = 0$ b) $2x^2 + x - 1 = 0$
c) $5x^2 + 3x + 3 = 0$ d) $7x^2 - 5x + 1 = 0$

8 Which of the following equations have equal roots?
a) $25x^2 - 10x + 2 = 0$ b) $9x^2 - 12x + 4 = 0$
c) $4x^2 + 4x + 1 = 0$ d) $2x^2 + 4x + 2 = 0$

9 The roots of the equation $x^2 - 7x + 5 = 0$ are α and β. Find the value of:

a) $\alpha + \beta$ b) $\alpha\beta$ c) $\alpha^2 + \beta^2$ d) $\dfrac{1}{\alpha} + \dfrac{1}{\beta}$ e) $\dfrac{\alpha}{\beta} + \dfrac{\beta}{\alpha}$

10 The roots of the equation $x^2 + 8x + 4 = 0$ are α and β. Find the value of:

a) $\alpha^2 + \beta^2$ b) $\dfrac{1}{\alpha^2} + \dfrac{1}{\beta^2}$ c) $(\alpha + \beta)^3 - 3\alpha\beta(\alpha + \beta)$ d) $\alpha^3 + \beta^3$

11 The roots of the equation $3x^2 + 7x - 4 = 0$ are α and β ($\alpha > \beta$). Find the value of:

a) $\dfrac{1}{\alpha} + \dfrac{1}{\beta}$ b) $\dfrac{\alpha}{\beta} + \dfrac{\beta}{\alpha}$ c) $\alpha^2 + \beta^2$ d) $\alpha - \beta$ e) $\alpha^2 - \beta^2$

12 The roots of the equation $x^2 + px + q = 0$ are α and β. Find the equations whose roots are:

a) $\dfrac{1}{\alpha}, \dfrac{1}{\beta}$ b) α^2, β^2 c) $\dfrac{\alpha}{\beta}, \dfrac{\beta}{\alpha}$

d) $\alpha + 1, \beta + 1$ e) $\alpha - \beta, \beta - \alpha$ f) $\dfrac{1}{\alpha^2}, \dfrac{1}{\beta^2}$

13 Without solving the equation, write down the equation whose roots are the reciprocals of the roots of $4x^2 + 2x - 1 = 0$.

14 Without solving the equation, write down the equation whose roots are the squares of the roots of $5x^2 - 3x - 2 = 0$.

15 Without solving the equation, write down the equation whose roots are double those of the equation $2x^2 + 7x + 1 = 0$.

16 For what value of k are the roots of $4x^2 + (k + 2)x - 5 = 0$ equal and opposite?

17 One root of the equation $4x^2 + 7x + c = 0$ is equal to the reciprocal of the other. Find the value of c.

18 One root of the equation $4x^2 + bx + 3 = 0$ is three times the other. Find the values of b.

19 The equation $ax^2 + bx + c = 0$ ($a \neq 0$) has roots α and β. Write down expressions for $\alpha + \beta$ and $\alpha\beta$ in terms of a, b and c. Find the quadratic equation whose roots are $\alpha^2\beta$ and $\beta^2\alpha$.

20 The roots of the quadratic equation $ax^2 + bx + c = 0$ are α and 3α. By considering the sum and product of the roots, show that $3b^2 = 16ac$.

Partial Fractions

3

Type 1 Linear factors in denominator,

e.g. $$\frac{4}{(x+3)(x-2)} \equiv \frac{A}{x+3} + \frac{B}{x-2}$$

Type 2 Repeated linear factors in denominator,

e.g. $$\frac{3}{(x+4)^2(x-3)} \equiv \frac{A}{x+4} + \frac{B}{(x+4)^2} + \frac{C}{x-3}$$

Type 3 Quadratic factors that will not factorise in the denominator,

e.g. $$\frac{1}{(x^2+1)(x-5)} \equiv \frac{Ax+B}{x^2+1} + \frac{C}{x-5}$$

Type 4 The degree of the top is equal to or greater than the degree of the bottom,

e.g. $$\frac{2x^2+x+1}{(x+1)(x-2)} \equiv \frac{2x^2-2x-4+3x+5}{x^2-x-2}$$

$$\equiv 2 + \frac{3x+5}{(x+1)(x-2)}$$

i.e. $$\frac{2x^2+x+1}{(x+1)(x-2)} \equiv 2 + \frac{A}{x+1} + \frac{B}{x-2}$$

EXERCISE 3

1 Find A and B if $(x+3)(x+2) \equiv x^2 + Ax + B$.

2 Find A and B if $A(x-2)(x+4) + B(x-3)(x+3) \equiv 2x+1$.

3 Find A, B and C if $Ax(x+3) + B(x-3) + C \equiv x^2$.

4 Find A and B if $\dfrac{2x+5}{(x+2)(x+3)} \equiv \dfrac{A}{x+2} + \dfrac{B}{x+3}$.

5 Find A, B and C if $\dfrac{4x+6}{(x+1)(x+2)(x+3)} \equiv \dfrac{A}{x+1} + \dfrac{B}{x+2} + \dfrac{C}{x+3}$.

6 Find A and B if $\dfrac{x+3}{(x+1)^2} \equiv \dfrac{A}{x+1} + \dfrac{B}{(x+1)^2}$.

7 Find A, B and C if $\dfrac{6x^2-5x-19}{(x-2)^2(x+3)} \equiv \dfrac{A}{x-2} + \dfrac{B}{(x-2)^2} + \dfrac{C}{x+3}$.

8 Find A and B if $\dfrac{4x-3}{(x^2+3)(x+4)} \equiv \dfrac{Ax}{x^2+3} + \dfrac{B}{x+4}$.

Express in partial fractions:

9 $\dfrac{5x-8}{x(x-4)}$

10 $\dfrac{3x-14}{(x-3)(x-4)}$

8

11 $\dfrac{3x^2 - 11x + 4}{(x-2)(x-3)(x-4)}$

12 $\dfrac{4x^2 - 15x - 1}{(x-1)(x+2)(x-3)}$

13 $\dfrac{1}{(x^2 - 1)(x-1)^2}$ (Factorise the quadratic factor in the denominator first.)

14 $\dfrac{7x + 8}{(x+2)(x^2 - 1)}$

15 $\dfrac{4x - 2}{(x^2 + 1)(x+2)}$

16 $\dfrac{x}{x^3 + 1}$

17 $\dfrac{11x + 5}{(x-3)(2x^2 + 1)}$

18 $\dfrac{x^2 + 3x - 1}{(x+2)(x+3)}$

19 $\dfrac{2x^2 + 1}{(x-1)(x+2)}$

20 $\dfrac{10x + 16}{(x^2 - 4)(x+1)}$

21 $\dfrac{1}{x^2 - a^2}$

22 $\dfrac{x^2 + 1}{x^2 - 1}$

23 $\dfrac{x^3 - 5x^2 + 10}{(x-2)(x-4)}$

24 $\dfrac{50}{(x+1)^2(x-4)}$

25 $\dfrac{2x^3 - x^2 - 13x - 13}{(x+1)^3(x-2)}$

Surds and Surd Equations

4

EXERCISE 4

1 Express as the square root of a single number:

a) $2\sqrt{2}$ b) $3\sqrt{3}$ c) $2\sqrt{5}$ d) $3\sqrt{7}$

2 Simplify:

a) $\sqrt{12}$ b) $\sqrt{18}$ c) $\sqrt{50}$ d) $\sqrt{45}$

3 Express as simply as possible:

a) $\sqrt{3} \times \sqrt{6}$ b) $2\sqrt{50}$ c) $\sqrt{2} \times \sqrt{3} \times \sqrt{6}$ d) $\sqrt{3} \times \sqrt{12}$

4 Rationalise the denominators and simplify:

a) $\dfrac{1}{\sqrt{2}}$ b) $\dfrac{3}{\sqrt{3}}$ c) $\dfrac{2}{\sqrt{8}}$ d) $\dfrac{25}{\sqrt{5}}$

e) $\dfrac{1}{1 + \sqrt{2}}$ f) $\dfrac{2}{\sqrt{3} - 1}$ g) $\dfrac{2}{\sqrt{5} + 1}$ h) $\dfrac{14}{2\sqrt{2} - 1}$

i) $\dfrac{11}{2\sqrt{3} + 1}$ j) $\dfrac{1}{\sqrt{3} - 1} + \dfrac{1}{\sqrt{3} + 1}$ k) $\dfrac{2}{\sqrt{2} + 1} - \dfrac{2}{\sqrt{2} - 1}$

5 Simplify:

a) $\sqrt{2}(2 - \sqrt{2})$

b) $\sqrt{3} + \sqrt{27}$

c) $(\sqrt{3} + 1)(\sqrt{3} - 1)$

d) $(\sqrt{2} + 1)(\sqrt{2} - 1)$

e) $(3\sqrt{3} + 1)(3\sqrt{3} - 1)$

f) $(3\sqrt{2} - 1)^2$

g) $(2\sqrt{3} - \sqrt{6})(1 + \sqrt{2})$

h) $(\sqrt{5} - \sqrt{10})(\sqrt{5} + \sqrt{20})$

i) $(2\sqrt{3} - \sqrt{5})^2$

6 Simplify:

a) $\dfrac{2 + \sqrt{2}}{\sqrt{2}}$

b) $\dfrac{3 + \sqrt{6}}{\sqrt{3}}$

c) $\dfrac{\sqrt{12} + 3}{\sqrt{3}}$

d) $2 - \dfrac{1}{\sqrt{2}} + \sqrt{2}$

e) $\dfrac{6}{\sqrt{2}} + \sqrt{2} - 4$

f) $\dfrac{1}{\sqrt{5}} + \dfrac{\sqrt{5}}{2}$

g) $\dfrac{\sqrt{2} + 1}{\sqrt{2} - 1}$

h) $\dfrac{3\sqrt{3} - 1}{\sqrt{3} + 1}$

i) $\dfrac{2 - \sqrt{5}}{\sqrt{5} + 1}$

Solve the following equations:

7 $\sqrt{5x + 1} = 4$

8 $\sqrt{4x - 3} = x$

9 $\sqrt{2x + 1} - \sqrt{x} = 1$

10 $\sqrt{5x + 1} - \sqrt{2x - 2} = 2$

11 $\sqrt{x + 6} - \sqrt{4 - x} = 2$

12 $\sqrt{5x + 6} - \sqrt{2x} = 2$

13 $\sqrt{3x + 4} + 2 = \sqrt{7x}$

14 $\sqrt{5x} - \sqrt{x - 1} = \sqrt{2x - 1}$

15 $\sqrt{4x + 1} - 3 = \sqrt{x - 2}$

16 $\sqrt{7 - 6x} - \sqrt{x + 12} = \sqrt{1 - x}$

Arithmetic Progressions

5

For an arithmetic progression (AP) with first term a and common difference d:

$$T_n = a + (n - 1)d$$

$$S_n = \frac{n}{2}[2a + (n - 1)d]$$

EXERCISE 5

1 The 6th term of an AP is 13 and the 9th term is 19. Find (a) the 12th term (b) the sum to 20 terms.

2 Find the 9th and 15th terms in an AP with a first term of 8 and a common difference of -2.

3 The sum of the first n terms of an AP is $n^2 + 3n$. Show that the terms of the series are in arithmetic progression. Find the 10th term.

4 The sum of the first n terms of an AP is $\dfrac{n}{4}(3n + 45)$. Write down the first four terms of this series.

5 In an arithmetic progression the sum of the first term and the 5th term is zero and the 13th term is 20. Find the first term and the common difference, and hence show that the 7th term is twice the 5th term.

6 How many terms of the AP $3 + 9 + 15 + \ldots$ are needed to make a sum of 2187? Find the least number of terms of this series needed to make a sum exceeding 1000.

7 The 2nd term of an AP is 13, and the 6th term is 21. Find the common difference, the first term and the sum of the first 20 terms.

8 Find the sum of the first 100 odd numbers.

9 Find the sum of the odd numbers that are exactly divisible by 3, that lie between 200 and 400.

10 How many whole numbers are there between 300 and 500 that are exactly divisible by 7? What is the sum of those numbers?

11 Show that the sum of the first n integers is $\dfrac{n}{2}(n + 1)$.

12 The sum of the first n even integers is 2550. Find n.

13 The 19th term of an AP is 17 and the sum of the first 14 terms is $157\frac{1}{2}$. Find the first term and the common difference.

14 The 6th term of an AP is 2.6 and the sum of the 10th and 11th terms is 7. Find the first term, the common difference and the sum of the first 20 terms.

15 The 3rd term of an AP is 2 and the 6th term is -7. Find the sum from the 10th term to the 20th term inclusive.

16 The 3rd and 4th terms of an AP are respectively $\log_e ar^2$ and $\log_e ar^3$. Find the first term and the common difference. Show that the sum of n terms is $\dfrac{n}{2}\log_e(a^2 r^{n-1})$.

If $a = 2$ and $r = 2$, use this result to find the smallest value of n such that the sum exceeds 25.

17 Find the arithmetic mean of 8 and 32.

11

18 Insert three arithmetic means between 8 and 20.

19 Insert two arithmetic means between $3a - 3b$ and $3a + 3b$.

20 Insert three arithmetic means between a and $5a - 4b$.

Geometric Progressions

6

If the first term of a geometric progression (GP) is a and the common ratio is r then the nth term is given by:

$$T_n = ar^{n-1}$$

and the sum to n terms by:

$$S_n = \frac{a(r^n - 1)}{r - 1} \quad (r > 1)$$

$$S_n = \frac{a(1 - r^n)}{1 - r} \quad (r < 1)$$

$$S_\infty = \frac{a}{1 - r} \quad (|r| < 1)$$

EXERCISE 6

1 Find the sum of the first 40 terms of the geometric progression 1, -2, $+4$.

2 The 4th term of a GP is 20 and the 7th term is 160. Find the common ratio, the first term and the sum of the first six terms.

3 Hilary made a New Year's resolution to save. At the end of the first month she saved £1, at the end of the second £2, at the end of the third £4, and so on. How much did she aim to save at the end of December? How much would she have saved during the whole of the year?

4 The 7th term of a GP is $\frac{1}{4}$ and the 9th term is $\frac{1}{16}$. Find the common ratio and the first term.

5 Three numbers are in GP. The sum of the first two is 9 and sum of the second two is 18. Find the three numbers.

6 The sum of n terms of a GP is $8(2^n - 1)$. Find the first three terms.

7 In a GP the first term is 1 and the last term is 1024. How many terms are there if their sum is 2047?

8 The sum of the first two terms of a GP is 12 and the third term is 1. Find:
a) the two possible values of the common ratio
b) the sum to infinity of the series with the positive common ratio.

9 A GP with first term a and common ratio r has $2n$ terms. Show that the sum of this progression is $\dfrac{a(r^{2n} - 1)}{r - 1}$, and that the sum of the 1st, 3rd, 5th, ... $(2n - 1)$th terms (i.e. the odd terms) is $\dfrac{a(r^{2n} - 1)}{r^2 - 1}$. Hence show that the sum of the even terms is r times the sum of the odd terms.

10 a) Prove that $\log_b a = \dfrac{\log_c a}{\log_c b}$.
b) Show that $\log_3 x + \log_9 x + \log_{81} x + \ldots$ is a geometric progression with a common ratio of $\frac{1}{2}$.

What is the sum to infinity of this progression?

11 A ball is dropped from a height of 10 m on to a hard surface. After each impact it rises to one-third of the height from which it fell. How far does it travel altogether before it comes to rest?

12 A farmer took his horse to the blacksmith for four new shoes. Each shoe required seven nails and the farmer agreed to pay 1p for the first nail, 2p for the second nail, 4p for the third nail, 8p for the fourth nail, and so on. What was the total cost of shoeing the horse?

13 Assuming that 8 people lived on the Earth in the time of Noah (4000 BC) and that the population doubled every 100 years, what would the population of the Earth be in AD 2000?

14 An investment company offers its investors 8% compound interest. Sophie Kay invests £P. What will this amount grow to after: (a) 1 year, (b) 2 years, (c) n years? How many years will it take before she has doubled her money?

15 Find the geometric mean of 6 and 54.

16 Insert three geometric means between 2 and 32.

The Remainder Theorem

If a polynomial $f(x)$ is divided by $(x - a)$ until the remainder no longer contains x, the remainder is $f(a)$.

It follows that if $f(a) = 0$, then $(x - a)$ is a factor.

EXERCISE 7

1 Find the remainder when the given functions are divided by the linear factors that follow them.

a) $x^3 + x^2 - x + 3, \quad x - 1$
b) $x^3 - 2x^2 - 3x + 1, \quad x + 2$
c) $2x^3 + 3x - 4, \quad x - 3$
d) $3x^3 - x^2 + 5, \quad x + 1$
e) $3x^3 - 8x^2 - 5x + 2, \quad x - 4$
f) $4x^3 + 2x^2 + 3x - 105, \quad x + 3$

2 Verify that:

a) $(x - 2)$ is a factor of $x^3 - 3x^2 - 4x + 12$
b) $(x - 1)$ is a factor of $x^3 - 3x^2 + 3x - 1$
c) $(x + 3)$ is a factor of $x^3 + 27$
d) $(x + 3)$ is a factor of $x^3 - 19x - 30$
e) $(x + 2)$ is a factor of $x^3 + 6x^2 + 11x + 6$
f) $(x - 4)$ is a factor of $x^3 - x^2 - 10x - 8$

3 Factorise the following expressions as far as possible.

a) $x^3 - 2x^2 - 5x + 6$ b) $x^3 + 5x^2 + 2x - 8$
c) $x^3 + 3x^2 - 10x - 24$ d) $x^3 - x^2 - 10x - 8$
e) $x^3 + 6x^2 + 11x + 6$ f) $x^4 - 16$
g) $x^3 - 27$ h) $x^3 + x^2 - x - 1$
i) $x^3 - 3x^2 + x - 3$ j) $x^4 - 3x^3 + 3x^2 - 3x + 2$
k) $2x^3 - 3x^2 - 8x - 3$ l) $2x^3 + 9x^2 + 7x - 6$

4 Find a if $(x - 3)$ is a factor of $x^3 + ax^2 + 11x - 6$.

5 Find b if $(x + 2)$ is a factor of $x^3 + 6x^2 + bx + 6$.

6 When $x^3 + 2x^2 + ax + 7$ is divided by $x - 2$ the remainder is 3. Find a.

7 When $2x^3 + x^2 + bx + 58$ is divided by $x + 3$ the remainder is 4. Find b.

8 Both $x + 2$ and $2x - 1$ are factors of $2x^3 + ax^2 + bx + 6$. Find a and b, and the third factor.

9 Both $(x-2)$ and $(x+2)$ are factors of $x^4 + ax^3 + bx^2 + 16x - 12$. Find a and b, and hence factorise the expression completely.

10 When $x^3 + ax^2 - x + 14$ is divided by $x - 3$ the remainder is 11. Find a. What is the remainder when the resulting expression is divided by $x + 2$?

11 One root of the equation $x^2 + ax - 3 = 0$ is 3. Find a and hence find the other root.

12 Use the remainder theorem to solve the equation $x^3 + 2x^2 - 9x - 18 = 0$.

13 Use the remainder theorem to solve the equation $x^3 - x^2 - 14x + 24 = 0$.

14 One root of the equation $x^4 + ax^3 + bx^2 + 16x - 12 = 0$ is 2 and another root is -2. Find the values of a and b. Use these values to find the other two roots of the given equation.

15 If $f(x) = (x-a)^2 \Phi(x)$ prove that $(x-a)$ is a factor of both $f(x)$ and $f'(x)$.

Radian Measure 8

A radian is the angle subtended at the centre of a circle by an arc length equal to the radius of the circle.

$$1 \text{ radian} \approx 57.296°$$

$$\pi \text{ radians} = 180°$$

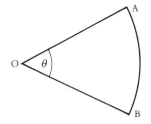

arc length AB $= r\theta$

area of sector OAB $= \frac{1}{2}r^2\theta$

EXERCISE 8

1 Convert to degrees the following angles given in radians.

a) $\frac{\pi}{2}$ b) $\frac{\pi}{4}$ c) $\frac{\pi}{3}$ d) $\frac{\pi}{6}$ e) $\frac{3\pi}{4}$ f) $\frac{2\pi}{3}$ g) 2π h) $\frac{7\pi}{2}$

2 Convert the following angles to radians, giving each answer as a multiple of π.

a) $45°$ b) $60°$ c) $120°$ d) $135°$ e) $270°$ f) $300°$
g) $480°$ h) $600°$ i) $720°$

15

3 Find, in radians, the angle subtended at the centre of a circle of radius 5 cm, by an arc:

a) 4 cm long,   b) 7 cm long.

4 Find, in radians, the angle subtended at the centre of a circle of radius 8 cm by an arc 5 cm long. What is the area of this sector?

5 The arc of a sector of a circle of radius 5 cm is 6 cm. What is the area of the sector?

6 The area of a sector of a circle of radius 8 cm is 40 cm². What is the angle contained by this sector?

7 A segment is cut off from a circle of radius 8 cm by a chord AB of length 10 cm. Find:

a) the size of the angle that the chord AB subtends at the centre of the circle (give your answer in radians).
b) the length of the minor arc AB,
c) the area of the segment.

8 The perimeter of a semicircle is 80 cm. Find the area of a sector subtending an angle of 60° at the centre.

9

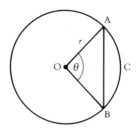

Find, in terms of r and θ, the area of:

a) the triangle OAB,
b) the sector OACB,
c) the minor segment ABC.

10

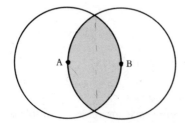

The figure shows two circles, centres A and B, each with a radius of 10 cm. Find the area common to both circles correct to the nearest whole number.

The Sine Rule and the Cosine Rule

THE SINE RULE

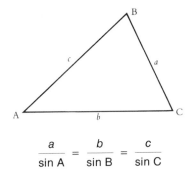

$$\frac{a}{\sin A} = \frac{b}{\sin B} = \frac{c}{\sin C}$$

THE COSINE RULE

$$a^2 = b^2 + c^2 - 2bc \cos A$$

$$b^2 = a^2 + c^2 - 2ac \cos B$$

$$c^2 = a^2 + b^2 - 2ab \cos C$$

EXERCISE 9

1 In triangle ABC, A = 43.4°, C = 73.5° and c = 12 cm. Find a and b.

2 In triangle ABC, a = 8.4 cm, b = 9.6 cm and B = 84.5°. Find A and C and the area of the triangle.

3 The side of a triangular field AB, of length 100 m, lies in a direction north-west. From A the corner C is in a direction S 30° W and from B, C is in a direction S 15° E. Find:
a) the distances of C from A and B,
b) the area of the field in square metres.

4 A boat is sailing directly towards the foot of a cliff. The angle of elevation of a point on the top of the cliff, and directly ahead of the boat, increases from 8° to 12° as the boat sails 100 m. Find:
a) the original distance from the boat to the point on the top of the cliff,
b) the height of the cliff.

5 In triangle ABC, a = 5.6 cm, b = 7.8 cm and C = 54°. Find c and A.

6 In triangle ABC, a = 12 cm, b = 13.5 cm and c = 18 cm. Find the size of the smallest angle in the triangle.

7 The lengths of the three sides of a triangle are 8.5 cm, 6.8 cm and 9.4 cm. Find:
a) the size of the largest angle in the triangle,
b) the area of the triangle.

8 The area of triangle ABC is 65 cm². If $b = 15$ cm and $c = 12$ cm, find A and a.

9 Find the angle between the hands of a clock at 12.15. If the hour hand is 8 cm long and the minute hand is 10 cm long, how far apart are the tips of the hands at this time?

10 A coaster leaves a port P and sails 5 nautical miles on a bearing of 052° followed by 6 nautical miles on a bearing of 205° to arrive at the next port Q. Find the distance and bearing of P from Q.

11

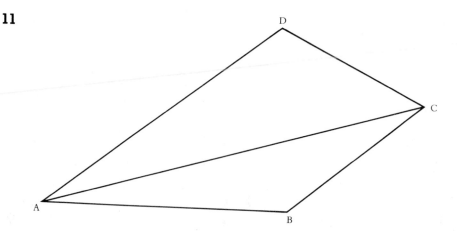

The diagram shows a field in the form of a quadrilateral ABCD. From A, B is 500 m due east and D is 600 m on a bearing of 062°. From B, C is 300 m on a bearing of 075°. Find:
a) the distance and bearing of C from A,
b) the length of the side CD,
c) the area of the field i) in m² ii) in hectares

giving your answers correct to three significant figures.

12 A, B, C and D represent four villages, B being due south of A. The distances AB, BC and CA are respectively 12 km, 5.6 km and 14.2 km, C being on the eastern side of AB. From A, D is 13.6 km on a bearing of 102°. Find:
a) angle BAC,
b) angle CAD,
c) the distance and bearing of D from C.

Trigonometric Identities

10

USEFUL FACTS

$$\tan A = \frac{\sin A}{\cos A} \qquad\qquad \cot A = \frac{\cos A}{\sin A}$$

$$\sec A = \frac{1}{\cos A} \qquad\qquad \operatorname{cosec} A = \frac{1}{\sin A}$$

$$1 + \tan^2 A = \sec^2 A \qquad\qquad 1 + \cot^2 A = \operatorname{cosec}^2 A$$

$$\sin^2 A + \cos^2 A = 1$$

$$\sin (A + B) = \sin A \cos B + \cos A \sin B$$
$$\sin (A - B) = \sin A \cos B - \cos A \sin B$$
$$\cos (A + B) = \cos A \cos B - \sin A \sin B$$
$$\cos (A - B) = \cos A \cos B + \sin A \sin B$$

$$\sin 2A = 2 \sin A \cos A \qquad\qquad \begin{aligned} \cos 2A &= \cos^2 A - \sin^2 A \\ &= 2 \cos^2 A - 1 \\ &= 1 - 2 \sin^2 A \end{aligned}$$

$$\tan 2A = \frac{2 \tan A}{1 - \tan^2 A}$$

$$\tan (A + B) = \frac{\tan A + \tan B}{1 - \tan A \tan B} \qquad \tan (A - B) = \frac{\tan A - \tan B}{1 + \tan A \tan B}$$

$$\cos^2 A = \tfrac{1}{2}(1 + \cos 2A) \qquad\qquad \sin^2 A = \tfrac{1}{2}(1 - \cos 2A)$$

$$\sin (A + B) \sin (A - B) = \sin^2 A - \sin^2 B$$

$$\sin 3A = 3 \sin A - 4 \sin^3 A \qquad\qquad \cos 3A = 4 \cos^3 A - 3 \cos A$$

$$\tan 3A = \frac{3 \tan A - \tan^3 A}{1 - 3 \tan^2 A}$$

If $t = \tan \tfrac{1}{2} A$

$$\sin A = \frac{2t}{1 + t^2}, \quad \cos A = \frac{1 - t^2}{1 + t^2}, \quad \tan A = \frac{2t}{1 - t^2}$$

$$\sin P + \sin Q = 2 \sin \frac{P + Q}{2} \cos \frac{P - Q}{2}$$

$$\sin P - \sin Q = 2 \cos \frac{P + Q}{2} \sin \frac{P - Q}{2}$$

$$\cos P + \cos Q = 2 \cos \frac{P + Q}{2} \cos \frac{P - Q}{2}$$

$$\cos P - \cos Q = -2 \sin \frac{P + Q}{2} \sin \frac{P - Q}{2}$$

19

EXERCISE 10

For questions 1 to 4, without using a calculator, show that:

1 $\sin 75° = \dfrac{\sqrt{3}+1}{2\sqrt{2}}$

2 $\sin 15° = \dfrac{\sqrt{3}-1}{2\sqrt{2}}$

3 $\cos 105° = \dfrac{1-\sqrt{3}}{2\sqrt{2}}$

4 $\sin 105° = \dfrac{1+\sqrt{3}}{2\sqrt{2}}$

5 If $\sin A = \frac{4}{5}$ and $\cos A = -\frac{5}{13}$, where A is acute and B is obtuse, find the exact value of: a) $\sin (A + B)$, b) $\cos (A - B)$.

6 If $\sin A = \frac{8}{17}$ and $\cos B = \frac{3}{5}$, where A and B are both acute, find the exact value of: a) $\sin (A - B)$, b) $\cos (A + B)$, c) $\tan (A + B)$.

7 If $\sin A = \frac{4}{5}$ and $\cos B = \frac{1}{3}$, where A is obtuse and B is acute, find in surd form the value of: a) $\sin (A + B)$, b) $\cos (A - B)$.

8 If $\cos A = \sin B = \tan C = \frac{1}{2}$, where A, B and C are respectively in the first, second and third quadrants, find the value of: a) $\sin (A + B + C)$, b) $\cos (A - B + C)$.

9 If $\sin (x - \beta) = \cos (x + \alpha)$ find an expression for $\tan x$ in terms of α and β.

10 Show that $\tan (A + 45°) = \dfrac{1 + \tan A}{1 - \tan A}$.

11 Show that $\tan (60° + \theta) = \dfrac{\sqrt{3} - \tan \theta}{1 - \sqrt{3}\tan \theta}$.

12 Prove that $\sin^2 A - \sin^2 B = \cos^2 B - \cos^2 A$.

13 Prove that $\operatorname{cosec} 2A = \cot A - \cot 2A$.

Prove that:

14 $\tan^2\theta - \sin^2\theta = \dfrac{\sin^4\theta}{\cos^2\theta}$

15 $\dfrac{\cos A + \sin A}{\cos A - \sin A} = \dfrac{1 + \sin 2A}{\cos 2A}$

16 $\dfrac{1}{1 + \tan \theta} = \dfrac{\cot \theta}{1 + \cot \theta}$

17 $\operatorname{cosec}^2 A \cos^2 A = \operatorname{cosec}^2 A - 1$

18 $\dfrac{1}{1 + \cos A} + \dfrac{1}{1 - \cos A} = 2\operatorname{cosec}^2 A$

19 $\dfrac{1}{\cos \alpha + \sin \alpha} + \dfrac{1}{\cos \alpha - \sin \alpha} = \dfrac{\tan 2\alpha}{\sin \alpha}$

20 $\dfrac{\sin \theta}{\sin \phi} + \dfrac{\cos \theta}{\cos \phi} = \dfrac{2\sin (\theta + \phi)}{\sin 2\phi}$

20

21 $\dfrac{1 - \cos 2x}{1 + \cos 2x} = \tan^2 x$

22 $(\sin A + \cos A)^2 + (\sin A - \cos A)^2 = 2$

23 $\dfrac{\cos \alpha}{1 + \sin \alpha} + \dfrac{1 + \sin \alpha}{\cos \alpha} = 2 \sec \alpha$

24 $(\cos \theta - \cos \phi)^2 + (\sin \theta - \sin \phi)^2 = 2[1 - \cos (\theta - \phi)]$

25 $\tan^2 \left(\frac{\pi}{4} - \theta\right) = \dfrac{1 - \sin 2\theta}{1 + \sin 2\theta}$

26 $\dfrac{\sin \theta + \sin 3\theta + \sin 5\theta}{\cos \theta + \cos 3\theta + \cos 5\theta} = \tan 3\theta$

27 $\dfrac{\sin A + \sin 3A + \sin 5A + \sin 7A}{\cos A + \cos 3A + \cos 5A + \cos 7A} = \tan 4A$

28 Eliminate θ from the pair of equations:
$$x = a \cos \theta, \quad y = b \sin \theta$$

29 Eliminate θ from the pair of equations:
$$x = a \operatorname{cosec} \theta, \quad y = b \cot \theta$$

30 If $x = a \cos \theta + b \sin \theta$ and $y = a \sin \theta - b \cos \theta$ show that $x^2 + y^2 = a^2 + b^2$.

Trigonometric Equations

11

EXERCISE 11

In questions 1 to 15 solve the equations for $0 \leqslant \theta \leqslant 360°$.

1 $\sin \theta = \frac{1}{2}$ **2** $\cos \theta = 0.8$

3 $\tan \theta = 1$ **4** $\tan \theta = 2.5$

5 $\cos \theta = -0.33$ **6** $\sin \theta = -\frac{3}{4}$

7 $\sin 2\theta = \frac{1}{2}$ **8** $\cos 2\theta = -\frac{1}{2}$

9 $\sin 3\theta = -0.72$ **10** $\tan 3\theta = 1$

11 $(2 \sin \theta - 1)(3 \sin \theta + 2) = 0$ **12** $4 \cos^2 \theta = 1$

13 $(\tan \theta - 2)(3 \tan \theta + 1) = 0$ **14** $6 \cos^2 \theta + 5 \cos \theta + 1 = 0$

15 $\sin^2 \theta - \cos^2 \theta = 1$

21

In questions 16 to 20 solve the equations for $-180 \leqslant \theta \leqslant 180°$.

16 $2 \sin^2\theta - \cos \theta - 1 = 0$ **17** $\cos 2\theta + \sin \theta = 0$

18 $\tan 2\theta + \tan \theta = 0$ **19** $\cos 2\theta - 3 \cos \theta - 2 = 0$

20 $2 \tan \theta - 2 \cot \theta = 3$

In questions 21 to 26 solve the equations for $0 \leqslant \theta \leqslant 2\pi$.

21 $\tan 2\theta = 1$ **22** $\sin \left(\theta - \frac{\pi}{4}\right) = 1$

23 $\cos \left(\theta + \frac{\pi}{4}\right) = \frac{1}{2}$ **24** $\tan \left(2\theta - \frac{\pi}{2}\right) = -1$

25 $\cos 2\theta + \cos \theta = 0$ **26** $2 \sin^2\theta + \cos \theta - 1 = 0$

27 Given that $\sin (\theta - 30°) + \cos (\theta + 45°) = 0$ show that $\tan \theta = \dfrac{\sqrt{2} - 2}{\sqrt{6} - 2}$ and hence solve the given equation for $0 \leqslant \theta \leqslant 180°$.

28 Show that $\sin \left(\frac{\pi}{4} + \theta\right) = \frac{1}{\sqrt 2} (\cos \theta + \sin \theta)$ and write down a similar expression for $\sin \left(\frac{\pi}{4} - \theta\right)$. Hence show that

$$\sin \left(\tfrac{\pi}{4} + \theta\right) + \sin \left(\tfrac{\pi}{4} - \theta\right) = \sqrt{2} \cos \theta$$

and that $\sin \left(\tfrac{\pi}{4} + \theta\right) - \sin \left(\tfrac{\pi}{4} - \theta\right) = \sqrt{2} \sin \theta$

Solve the equation $\sin \left(\frac{\pi}{4} + \theta\right) - \sin \left(\frac{\pi}{4} - \theta\right) = \frac{1}{2}$ for $0 \leqslant \theta \leqslant \pi$.

The Equation $a \sin \theta \pm b \cos \theta = c$ **12**

To solve the equation $a \sin \theta + b \cos \theta = c$ divide throughout by $\sqrt{a^2 + b^2}$ and introduce the acute angle α such that $\tan \alpha = \dfrac{b}{a}$, then

$$\sin (\theta + \alpha) = \frac{c}{\sqrt{a^2 + b^2}}$$

The equation $a \sin \theta - b \cos \theta = c$ is solved in a similar way.

EXERCISE 12

1 Express $3 \sin \theta + 4 \cos \theta$ in the form $R \sin (\theta + \alpha)$.

2 Express $3 \sin \theta - 2 \cos \theta$ in the form $R \sin (\theta - \alpha)$.

3 Express $2 \cos \theta + \sin \theta$ in the form $R \cos (\theta - \alpha)$.

4 Express $4 \sin \theta + 5 \cos \theta$ in the form $R \sin (\theta + \alpha)$.

5 Express $2 \cos \theta + 3 \sin \theta$ in the form $R \cos (\theta - \alpha)$.

6 Express $2 \cos \theta - \sin \theta$ in the form $R \cos (\theta + \alpha)$.

7 Express $3 \sin 2\theta + 4 \cos 2\theta$ in the form $R \sin (2\theta + \alpha)$. Hence find the maximum value of $3 \sin 2\theta + 4 \cos 2\theta$. What acute value of θ gives this maximum value?

8 Express $3 \cos \theta - 2 \sin \theta$ in the form $R \cos (\theta + \alpha)$. Hence find the minimum value of this expression. Find the smallest positive value of θ for which this minimum value occurs.

9 Solve the equation $3 \sin \theta + 4 \cos \theta = 2$ for $0° \leqslant \theta \leqslant 360°$.

10 Solve the equation $2 \sin \theta - \cos \theta = 1$ for $0° \leqslant \theta \leqslant 360°$.

11 Solve the equation $3 \cos \theta + 5 \sin \theta = 4$ for $0° \leqslant \theta \leqslant 360°$.

12 Find R and α if $\sqrt{3} \cos \theta + \sin \theta = R \cos (\theta - \alpha)$. Hence solve the equation $\sqrt{3} \cos \theta + \sin \theta = 2$ for $-180° \leqslant \theta \leqslant 180°$.

13 Find the maximum and minimum values of $5 \sin \theta + 12 \cos \theta$, and the values of θ between $0°$ and $360°$, giving these values.

14 Show that $\cos \theta + \sqrt{3} \sin \theta$ can be written as $2 \cos (\theta - 60°)$ and as $2 \sin (\theta + 30°)$. Hence solve the equation $\cos \theta + \sqrt{3} \sin \theta = 2$ for $0° \leqslant \theta \leqslant 360°$.

15 Find the values of θ between $0°$ and $360°$ that satisfy the equation $\sin 2\theta + 2 \cos 2\theta = 1$.

16 Solve, for values of θ between $0°$ and $360°$, the equation $\sec \theta - 3 \tan \theta = 2$.

Trigonometry in Three Dimensions

13

EXERCISE 13

1

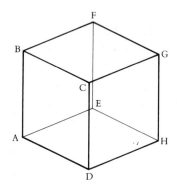

ABCDEFGH is a cube of side *a*. Find:

a) the angle between the line BD and the line AD,

b) the angle between the line BH and the plane ADHE,

c) the angle between the plane GED and the plane ADHE.

2

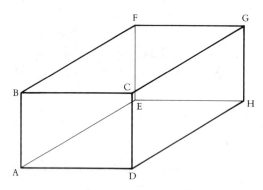

The sketch shows a cuboid in which AB = 2 cm, AE = 4 cm and AD = 3 cm.

a) Find the angles between:
   i)   AH and DE,
   ii) AG and AH,
   iii) the line BH and the plane ADHE,
   iv) the planes AFGD and AEHD,
   v)   BH and FD.

b) Find the lengths of the sides and the sizes of the angles in triangle BGD.

3

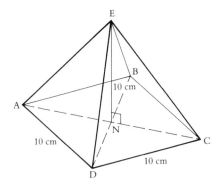

ABCDE is a square pyramid. The square base is of side 10 cm and E is 10 cm vertically above N, the point of intersection of the diagonals of the square. Calculate:

a) the angle between AE and the base,
b) the length of the edge AE,
c) the perpendicular distance from E to AB,
d) the angle between the planes AEB and ABCD.

4

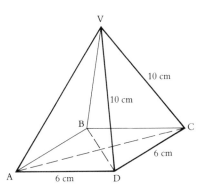

VABCD is a square pyramid. The square base has side 6 cm and the sloping edges are of length 10 cm. Find:

a) the angle between a sloping edge and the base,
b) the angle between one of the sloping faces and the base,
c) the height of the vertex, V, above the base,
d) the angle between any two adjacent sloping faces.

5

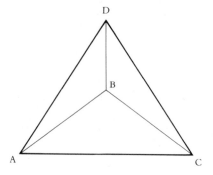

ABCD is a regular tetrahedron, each side being 8 cm. Find:

a) the shortest distance from D to AC,

b) the distance from D to the plane ABC,

c) the angle between AD and the plane ABC,

d) the angle between the plane ADB and the plane ABC.

6

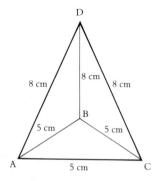

ABCD is a tetrahedron in which the base ABC is an equilateral triangle of side 5 cm and each sloping side is 8 cm long. Find:

a) the height of D above the base ABC,

b) the angle between DC and the base ABC,

c) the angle between the planes ABD and BCD,

d) the angle between the planes ABD and ABC.

The Straight Line **14**

The distance between the points $A(x_1, y_1)$ and $B(x_2, y_2)$ is given by:

$$AB = \sqrt{(x_2 - x_1)^2 + (y_2 - y_1)^2}$$

If $P(x, y)$ is the point that divides the join of the points $A(x_1, y_1)$ and $B(x_2, y_2)$ in the ratio $\lambda : \mu$,

$$x = \frac{\lambda x_2 + \mu x_1}{\lambda + \mu}, \quad y = \frac{\lambda y_2 + \mu y_1}{\lambda + \mu}$$

If $M(x, y)$ is the midpoint of AB:

$$x = \frac{x_1 + x_2}{2}, \quad y = \frac{y_1 + y_2}{2}$$

The gradient of AB is:

$$\frac{y_2 - y_1}{x_2 - x_1}$$

and the equation of the straight line through A and B is:

$$y - y_1 = \frac{y_2 - y_1}{x_2 - x_1}(x - x_1)$$

The equation $y = mx + c$ represents a straight line with gradient m and y-intercept c.

The equation of the straight line with gradient m passing through the point (x_1, y_1) is:

$$y - y_1 = m(x - x_1)$$

The angle (α) between the two straight lines with equations $y = m_1 x + c_1$ and $y = m_2 x + c_2$ is given by:

$$\tan \alpha = \pm \frac{m_1 - m_2}{1 + m_1 m_2}$$

The lines are perpendicular if $m_1 m_2 = -1$.

The area of the triangle with vertices $A(x_1, y_1)$, $B(x_2, y_2)$ and $C(x_3, y_3)$ is given by:

$$\triangle ABC = \pm \tfrac{1}{2}[x_1(y_2 - y_3) + x_2(y_3 - y_1) + x_3(y_1 - y_2)]$$

or

$$\triangle ABC = \frac{1}{2} \begin{vmatrix} x_1 & x_2 & x_3 \\ y_1 & y_2 & y_3 \\ 1 & 1 & 1 \end{vmatrix}$$

The straight line passing through the point $A(a, 0)$ on the x-axis, and $B(0, b)$ on the y-axis, has equation:

$$\frac{x}{a} + \frac{y}{b} = 1$$

The distance of the point (x_1, y_1) from the straight line with equation $Ax + By + C = 0$ is:

$$\pm \frac{Ax_1 + By_1 + C}{\sqrt{A^2 + B^2}}$$

EXERCISE 14

1 Find the distance between the points (2, 2) and (6, 5).

2 Find the distance between the points A (−5, −2) and B (7, 3). Write down the coordinates of M, the midpoint of AB.

3 Find the coordinates of the point P that divides the join of the points A (3, 1) and B (9, 4) in the ratio 2:1.

4 Find the coordinates of the point that divides the join of the points (−7, 3) and (8, −2) in the ratio 2:3.

5 A, B and C are the points (−5, 4), (4, 7) and (6, 1) respectively. Show that AB is perpendicular to BC and find the coordinates of D if ABCD is a rectangle.

6 Show that the points A (−6, −5), B (−3, −1) and C (3, 7) are collinear. Does the point D (−1, 3) lie above, on or below the straight line ABC?

7 Show that the points (−2, 5), (7, 2), (3, −2) and (−6, 1) are the vertices of a parallelogram. What is the length of the shorter diagonal?

8 A, B, C and D are the points (2, −2), (6, −4), (5, 4) and (3, 5). Show that the acute angle between AC and BD is 45°.

9 In triangle OAB the coordinates of A and B are respectively (11, 7) and (10, −6).
   a) Find: i) the acute angle between OA and OB,
          ii) the area of triangle OAB,
          iii) the coordinates of M, the midpoint of OB.
   b) Prove that: i) AM is perpendicular to OB,
               ii) triangle OAB is isosceles.

10 Find the equation of the perpendicular bisector of the join of the points A (−2, −3), B (8, 1). Show that this line passes through the point C (1, 4) and find the area of triangle ABC.

11 Find the area of the quadrilateral whose vertices are (−2, 1), (11, −4), (7, 5) and (−3, −5).

12 Find the equation of the straight line:
   a) passing through the points (−2, −3), (7, 2),
   b) with gradient $-\frac{1}{2}$ and passing through the point (5, −2),
   c) passing through the point (5, 2) and perpendicular to the line whose equation is $2x + 3y − 4 = 0$,
   d) passing through the point (−2, 4) and parallel to the line whose equation is $4x − 3y + 5 = 0$,
   e) making an intercept of 3 on the x-axis and −4 on the y-axis.

13 Show that the distance of the point $(2, 1)$ from the straight line with equation $3x + 4y + 10 = 0$ is equal to the distance of the point $(3, -2)$ from the straight line with equation $12x + 5y + 26 = 0$.

14 Show that the angle between the lines $x + y = 4$ and $3x - y - 8 = 0$ is equal to the angle between the lines $x + 3y - 9 = 0$ and $x - y - 4 = 0$.

15 Find the equation of the locus of a point P which moves in such a way that:
a) it is always twice as far from the x-axis as it is from the y-axis,
b) the sum of its distance from the x-axis and its distance from the y-axis is 10,
c) it is always the same distance from the point $(-3, 2)$ as it is from the point $(5, 3)$,
d) if A is the point $(1, 2)$ and B the point $(5, -2)$, the angle APB is always $90°$.

16 Show that the locus of a point P that is equidistant from the point $(-2, 3)$ and the line $2x + y - 4 = 0$ has equation $x^2 - 4xy + 4y^2 + 36x - 22y + 49 = 0$.

17 Find the ratio in which the straight line with equation $3x + 2y - 12 = 0$ divides the line joining the points A $(-4, 1)$ and B $(11, 6)$.

18 The vertices of triangle ABC are A $(-4, -1)$, B $(3, 5)$ and C $(8, 2)$. Find:
a) the equation of the altitude BN,
b) the coordinates of N,
c) the length of AC,
d) the area of triangle ABC.

19 The coordinates of the vertices of a triangle ABC are A $(7, 1)$, B $(1, 8)$ and C $(1, -2)$. Find the coordinates of the orthocentre (the intersection of the altitudes) of this triangle.

20 ABC is a triangle in which A is the point $(1, -1)$, B $(5, 7)$ and C $(9, -3)$. Find the coordinates of the centroid of this triangle, i.e. the coordinates of the intersection of the medians.

21 Find the coordinates of the incentre (the intersection of the bisectors of the angles) of a triangle with vertices $(3, -1)$, $(6, 5)$ and $(-5, 3)$.

Indices and Logarithms

15

USEFUL FACTS

$$\text{If } \log_b a = c \quad \text{then} \quad a = b^c$$

$$\log_c ab = \log_c a + \log_c b$$

$$\log_c \frac{a}{b} = \log_c a - \log_c b$$

$$\log_c a^n = n \log_c a$$

$$\log_b a = \frac{\log_c a}{\log_c b} = \frac{1}{\log_a b}$$

$$\log_a 1 = 0$$

EXERCISE 15

Find x if:

1 $3^{x-1} = 9$

2 $3^{2x-1} = 27$

3 $2^{3x} = 4^{x+1}$

4 $4^x = 6$

5 $2^{x-1} = 7$

6 $5^x = 3^{x+1}$

Find:

7 $\log_3 27$

8 $\log_4 64$

9 $\log_4 1$

10 $\log_9 3$

11 Prove that $\log 3 + \log 16 = \log 12 + \log 4$.

12 Simplify: a) $\log 6 + \log 12$,
b) $\log 12 - \log 6$,
c) $\frac{1}{2} \log 9 + \log 27$.

13 Simplify: a) $\dfrac{\log 4}{\log 2}$, b) $\dfrac{\log 27}{\log 3}$.

14 Express $\log_b a$ in terms of logarithms to the base c.

15 Prove that $\log_b a \times \log_c b \times \log_a c = 1$.

16 Prove that if $\log_a b = \log_b c = \log_c a$ then $a = b = c$.

17 Prove that $\log \dfrac{a}{b} + \log \dfrac{b}{c} + \log \dfrac{c}{a} = 0$.

18 If $\log_c a = b$ and $\log_b c = a$, prove that $\log_b a = ab$.

19 If $a^2 = bc$ prove that $2(\log_b a - \log_c a) = \log_b c - \log_c b$.

20 If $b^2 = \dfrac{a}{c}$ prove that $\log_b a - \log_b c = 2$.

21 Given that $a^2 + b^2 = 5ab$ prove that $2\log(a - b) - \log ab = \log 3$.

22 Find x if $\log_3 x = \log_9 (2x - 1)$.

23 Solve the equation $\log_3 x + \log_x 3 = 2$.

24 Solve the equation $2^{2x+3} - 7(2^{x+1}) - 15 = 0$.
(*Hint*: express the equation as a quadratic in 2^x.)

Log–log and Other Straight-line Graphs   **16**

Relationships connecting variables can be verified by drawing straight-line graphs.

A set of variables, x and y, is thought to be related by an equation of the form $y = ax^n$. It follows that $\log y = \log a + n \log x$.

If we plot $\log y$ against $\log x$ and show that the points lie approximately (or exactly) on a straight line then the supposition is verified. The gradient of the line gives the value of n and the y-intercept the value of $\log a$.

Similarly if it is thought that another set of values satisfies the equation $y = a + \dfrac{b}{x}$ we can verify this relationship if we obtain a straight line when we plot y against $\dfrac{1}{x}$.

EXERCISE 16

1 In a laboratory experiment two variables X and Y are known to be connected by a law of the form $Y = aX^n$ where a and n are constants. One student obtained the following results:

X	2.1	3.4	4.7	5.2	6.9	7.3
Y	34.4	120	280	435	758	880

Construct a new table showing $\log X$ and $\log Y$ and by plotting these new values on a graph show that the experimental data is a good approximation to the law. Which pair of results are obviously in error?

Use your graph to find the values of the constants a and n, giving each value correct to one decimal place.

2   In an experiment the following values of x and y were obtained:

x	1.2	2.6	3.4	4.2	6.5	8
y	2.08	3.87	4.79	5.67	8.05	9.50

It is known that the law connecting x and y is of the form $y = ax^n$ where a and n are constants. Plot values of $\log y$ against those of $\log x$ and hence show that the law is valid. Use your graph to determine the values of a and n.

3   Two variables p and q are known to be connected by the formula $pq^m = N$. In an experiment the following values of p and q were obtained:

p	1.5	2	3	4	5	6
q	5.05	4.22	3.27	2.73	2.38	2.12

By plotting the values of $\log p$ against $\log q$ show that the formula is valid. Take each log correct to two decimal places and take 2 cm as one-tenth of a unit on each axis.

For your graph write down the gradient of the resulting straight line together with the $\log p$ intercept. Hence determine the values of m and N.

4   Two variables D and V are connected by the formula $D^2 = KV^3$, corresponding values of which are given below:

D	1	2	2.4	3	3.5	4	4.2
V	1.13	1.79	2.02	2.34	2.60	2.84	2.92

Plot D^2 against V^3 to confirm the truth of this formula and use your graph to find the value of K.

5   Two variables P and Q are known to be connected by a law of the form $Q = aP^3 + b$, where a and b are constants.

In an experiment the following values of P and Q were obtained:

P	5.5	6.8	8	11.3	12	13.6	14
Q	650	870	1170	2560	2990	4170	4520

Draw a graph of P^3 against Q taking 6 cm $= 1000$ units on both axes and use your graph to evaluate the values of a and b. Find:
a) Q when $P = 10$,
b) P when $Q = 3500$.
(Give your answers correct to two significant figures.)

32

6 Two quantities x and y are known to be related by the law $y = a + \dfrac{b}{x^2}$ where a and b are constants. Values obtained in an experiment are given in the following table:

x	3.5	4.2	6	7.3	8.7
y	7.45	6.70	5.83	5.56	5.40

Draw a graph, plotting values of y against those of $\dfrac{1}{x^2}$. Choose scales which maximise the use of your graph paper.

Use your graph to estimate the values of a and b, and hence find the value of y when $x = 5$ and of x when $y = 8$.

7 In an experiment the following values of x and y were obtained:

x	2	3	4	5	6	7	8	9	10
y	10.40	20.71	33.78	49.36	67.30	87.46	109.7	134.1	160.5

It is known that the law relating these two quantities is of the form $y = ax^n$. Plot the values of $\log y$ against $\log x$ to show that the above law is true, and from your graph determine a and n, giving your values correct to one decimal place.

8 Two variables x and y are thought to be related by the formula:

$$e^y = ax^b$$

where a and b are constants. In an experiment the following pairs of values were obtained:

x	1	2	4	6	10
y	3.00	3.34	3.69	3.89	4.15

a) Express y in terms of a, b and x.
b) Draw a suitable straight-line graph and use it to determine values for a and b.
c) Use these values of a and b to find:
   i) the value of y when $x = 8$,
   ii) the value of x when $y = 3.5$.

Permutations and Combinations

$n!$ is short for $n \times (n - 1) \times (n - 2) \times \ldots \times 4 \times 3 \times 2 \times 1$.

For example $7! = 7 \times 6 \times 5 \times 4 \times 3 \times 2 \times 1 = 5040$.

The number of arrangements, or permutations, of n unlike things taken r at a time is ^nP_r where:

$$^nP_r = \frac{n!}{(n - r)!}$$

The number of combinations of n unlike things taken r at a time is ^nC_r where:

$$^nC_r = \frac{n!}{r!(n - r)!}$$

Note that $^nC_r = {}^nC_{n-r}$. For example $^{10}C_3 = {}^{10}C_7$.

The number of arrangements of n unlike things in a row if there are p alike of one kind, q alike of another kind, and so on is:

$$\frac{n!}{p!\,q!\ldots}$$

The number of ways of arranging n unlike things around a circle, considering clockwise and anticlockwise arrangements as different, is $(n - 1)!$

If no distinction is made between clockwise and anticlockwise (e.g. beads on a wire) the number of arrangements is $\dfrac{(n - 1)!}{2}$

EXERCISE 17

Evaluate:

1 $4!$

2 $8!$

3 $7! \times 3!$

4 $\dfrac{7!}{4!}$

5 $\dfrac{10!}{5!}$

6 $\dfrac{10!}{6!\,4!}$

7 $\dfrac{12!}{10!\,2!}$

8 $\dfrac{11!}{7!\,4!}$

9 $\dfrac{20!}{15!\,5!}$

10 Write in factorial form:

  a) $7 \times 6 \times 5 \times 4$

  b) $9 \times 8 \times 7$

  c) $12 \times 11 \times 10 \times 9$

  d) 7×3

  e) $\dfrac{9 \times 8 \times 7}{3 \times 2}$

11 Evaluate:

  a) $^6P_3$ b) $^7P_2$ c) $^{10}P_7$

12 Evaluate:

a) $^5C_2$ b) $^7C_4$ c) $^8C_6$

13 In how many different ways can eight books be arranged in a line on a shelf?

14 In how many different ways can ten books be arranged in a line on a shelf if two particular books must be next to each other?

15 In how many ways can the letters of the word HOUSE be arranged? How many of these arrangements start with the letter H? How many start with H and end with E?

16 Find the number of permutations of the letters of the word AFTERNOON.

17 Find the number of permutations of the letters of the word STATISTICS.

18 In how many different ways can eight people be arranged around a circular table?

19 In how many different ways can ten beads, each of a different colour, be arranged on a circular wire?

20 Ten questions are to be ordered in an exercise so that the easiest question comes first and the most difficult question comes last. The other eight questions are all of equal difficulty. In how many different ways can this be done?

21 How many even numbers of four digits can be formed from the figures 1, 2, 4, 8 if repetitions are allowed?

22 In how many ways can 3 books be chosen from a batch of 12?

23 The 11 players for a hockey team are to be chosen from a squad of 15. How many different team sheets can be written up if each player is willing to be selected in any position?

24 A manager has a squad of 22 players, two of whom are goalkeepers. In how many ways can he select a team of 11 players if all the other players can play in any position other than in goal?

25 Find the number of ways in which 10 adults can be divided into 2 equal groups if 2 particular adults must be in different groups.

26 There are 12 questions on an examination paper, 5 in section A and the remainder in section B. A candidate must attempt exactly 5 questions, at least two of which must be from each section. How many different choices does the candidate have?

27 A committee of 3 boys and 5 girls is to be chosen from 5 boys and 8 girls. In how many ways can this be done?

28 Ten points are marked on a circle. They are joined in pairs to give chords of the circle. How many different chords are possible?

29 What is the greatest number of points of intersection of 8 straight lines and 5 circles?

30 Three mixed pairs are to be chosen from 6 men and 7 women. In how many ways can this be done?

31 Sean wants to forecast the results of five football matches, each of which can be a win, a score draw, a no-score draw or a loss, for the home team. How many different forecasts are possible?

32 Esther takes four cups and saucers, each cup and saucer from a different set, and lays them on the table. In how many ways can she do this so that all the cups are on the wrong saucers?

The Binomial Theorem

18

$^{n}C_{r}$ or $\binom{n}{r}$ is short for $\dfrac{n!}{r!(n-r)!}$.

Hence, for example, $^{9}C_{5} = \dfrac{9!}{5!\,4!} = \dfrac{9 \times 8 \times 7 \times 6}{4 \times 3 \times 2 \times 1} = 126.$

If n is a positive integer:

$$(a + x)^{n} = {}^{n}C_{0}a^{n} + {}^{n}C_{1}a^{n-1}x + {}^{n}C_{2}a^{n-2}x^{2} + \ldots$$

$$\ldots + {}^{n}C_{r}a^{n-r}x^{r} + \ldots + {}^{n}C_{n}x^{n}$$

Note that the $(r + 1)$th term is $^{n}C_{r}a^{n-r}x^{r}$.

When only the first few terms are needed it is more convenient to write:

$$(a + x)^{n} = a^{n} + na^{n-1}x + \frac{n(n-1)}{2!}a^{n-2}x^{2} + \frac{n(n-1)(n-2)}{3!}a^{n-3}x^{3} + \ldots$$

If $a = 1$, this becomes:

$$(1 + x)^{n} = 1 + nx + \frac{n(n-1)}{2!}x^{2} + \frac{n(n-1)(n-2)}{3!}x^{3} + \ldots x^{n}$$

This result is extremely useful and is true for any value of x when n is a positive integer. If n is a negative integer, or a positive or negative fraction, the expansion is only valid if $|x| < 1$. For example

$$\frac{1}{1-x} = (1-x)^{-1} = 1 + (-1)(-x) + (-1)(-2)\frac{(-x)^2}{2!} + (-1)(-2)(-3)\frac{(-x)^3}{3!}$$

$$= 1 + x + x^2 + x^3 + \ldots$$

and

$$\sqrt{(1+x)} = (1+x)^{1/2} = 1 + \left(\frac{1}{2}\right)x + \left(\frac{1}{2}\right)\left(-\frac{1}{2}\right)\frac{x^2}{2!} + \left(\frac{1}{2}\right)\left(-\frac{1}{2}\right)\left(-\frac{3}{2}\right)\frac{x^3}{3!} + \ldots$$

$$= 1 + \frac{1}{2}x - \frac{1}{8}x^2 + \frac{1}{16}x^3 + \ldots$$

EXERCISE 18

Use the binomial theorem to expand the following expressions:

1 $(1 + 2x)^3$ **2** $(1 + 3x)^4$ **3** $(1 + x)^5$

4 $(1 - 2x)^4$ **5** $\left(1 - \frac{1}{2}x\right)^3$ **6** $\left(x + \frac{1}{x}\right)^3$

7 $(3x - 1)^3$ **8** $\left(x - \frac{2}{x}\right)^4$ **9** $\left(3 + \frac{x}{2}\right)^4$

Write down the first four terms in the binomial expansion of:

10 $\left(1 - \frac{x}{2}\right)^7$ **11** $(1 + 3x)^8$ **12** $\left(1 - \frac{5x}{2}\right)^{10}$

13 $(2 + x)^{10}$ **14** $(3 - 2x)^{12}$ **15** $\left(2 + \frac{x}{3}\right)^9$

16 Find the middle term in the binomial expansion of $\left(2 - \frac{x}{2}\right)^{12}$.

17 Find the two middle terms in the binomial expansion of $(3 + 2x)^{11}$.

18 Expand $(1 + x)(1 - 2x)^7$ as far as the term in x^4.

19 Find the values of a and b if

$$(1 - 2x)\left(1 - \frac{x}{2}\right)^8 = 1 + ax + bx^2 + \ldots$$

20 When $\left(1 - \frac{3x}{2}\right)^{12}$ is expanded in ascending powers of x the first three terms are $1 + ax + bx^2$. Find the values of a and b.

21 When $(1 - x)(1 + ax)^6$ is expanded as far as the term in x^2 the result is $1 + bx^2$. Find the values of a and b.

22 Expand $(1 - x + 2x^2)^6$ in ascending powers of x as far as x^3.

23 Find the coefficient of x^3 in the expansion of $(1 + x - x^2)^6$.

24 Find the coefficient of x^5 in the expansion of $(1 - 2x + 3x^2)^8$.

25 Find the term independent of x in the expansion of $\left(x - \dfrac{2}{x}\right)^8$.

26 Find the term independent of x in the expansion of $\left(x + \dfrac{1}{x^2}\right)^9$.

27 Use the binomial theorem to show that, if x is small,
$$\sqrt{\frac{1-x}{1+x}} \approx 1 - x + \frac{x^2}{2}$$
By putting $x = \frac{1}{8}$, show that $\sqrt{7} = \frac{339}{128}$.

28 Write down the coefficient of x^r in the binomial expansion of $(1 + ax)^n$. If this is c_r show that $\dfrac{c_{r+1}}{c_r} = \dfrac{a(n-r)}{r+1}$.

29 In the binomial expansion of $(2 + x)^n$ the coefficient of x^7 equals the coefficient of x^8. Find the value of n.

30 Write down the binomial expansion of $\left(1 + \dfrac{x}{2}\right)^n$ in ascending powers of x as far as x^3. Given that the coefficient of x^2 is equal to the coefficient of x^3 and that both coefficients are positive, find the value of n.

31 Use the binomial expansion to find $(1.0005)^{\frac{1}{5}}$ correct to eight decimal places.

32 Use the binomial expansion to find $(1.009)^{\frac{1}{3}}$ correct to six decimal places.

33 Expand $(1 - 3x)^{\frac{1}{2}}$ as a series of ascending powers of x, where $|x| < \frac{1}{3}$, up to and including the term in x^3. Express the coefficients in their simplest form.

Hence find the square root of 0.94 correct to five decimal places.

34 Find the first four terms in the expansion of $(1 - 2x)^{\frac{1}{3}}$ in ascending powers of x. State the set of values of x for which this expansion is valid. Use your result to find the cube root of 0.94 correct to five decimal places.

35 Evaluate the term that is independent of x in the binomial expansion of $\left(x - \dfrac{1}{x^2}\right)^{18}$.

36 Evaluate the term that is independent of x in the binomial expansion of $\left(x + \dfrac{1}{x}\right)^{16}$.

Inequalities 19

EXERCISE 19

Find the set of values of x for which:

1 $x^2 - 5x + 4 > 0$

2 $9 - x^2 > 0$

3 $15 + x - 2x^2 \geqslant 0$

4 $x^2 - 5x - 14 \leqslant 0$

5 $4x^2 - 9 < 0$

6 $12 - x - x^2 > 0$

7 $\dfrac{x-2}{x-3} > 2$

8 $\dfrac{x+6}{3x-1} < 2$

9 $\dfrac{2}{1-x} < \dfrac{1}{x+3}$

10 $(4x - 5)(2x - 3) > 4x$

11 $x - \dfrac{1}{x} < 1$

12 $2x - 4 > \dfrac{1}{x}$

In questions 13 to 16 find the set(s) of values of x for which the inequalities are satisfied.

13 $|x + 3| > |2x - 4|$

14 $|x| < |x^2 - 5x + 6|$

15 $|\cos x| > |\sin x|$ for $-\pi \leqslant x \leqslant \pi$

16 $2x + 5 > |x^2 - 3|$

17 If a and b are real show that:

a) $a^2 + b^2 \geqslant 2ab$

b) $(a + b)^2 \geqslant 4ab$

c) $\left| \dfrac{a}{b} + \dfrac{b}{a} \right| \geqslant 2$

18 If a, b and c are real show that $a^2 + b^2 + c^2 \geqslant ab + bc + ca$.

Simultaneous Equations

One Linear and One Second Degree

Find y (or x if it is easier) from the linear equation, substitute this value in the second-degree equation, simplify, factorise, and solve to give two values of the chosen variable. These variables are now substituted in the linear equation to complete the question.

EXERCISE 20a

Solve:

1 $x^2 + y^2 = 20$, $\quad y = x + 2$

2 $x^2 + y^2 = 25$, $\quad y = x + 1$

3 $y = 2x + 1$, $\quad x^2 + y^2 + 3x - 4y - 1 = 0$

4 $y = 5 - x$, $\quad x^2 + y^2 - 5x + 3y - 12 = 0$

5 $y = 5 + 2x$, $\quad x^2 + y^2 + 6x - 2y + 2 = 0$

6 $y = x + 1$, $\quad 9x^2 + y^2 - 3x - 4 = 0$

7 $2x^2 + y^2 + 9x - 9y - 45 = 0$, $\quad x - y + 6 = 0$

8 $3x^2 + xy - y^2 + 4x + 6y - 25 = 0$, $\quad 2x - y + 5 = 0$

9 $2x^2 + 9y^2 + 7x + 18y - 6 = 0$, $\quad x - 3y + 2 = 0$

10 $3x^2 - 16y^2 - x + 8y - 2 = 0$, $\quad 3x - 4y + 2 = 0$

11 $3x^2 - y^2 + 9x + 3y + 4 = 0$, $\quad y = 2x + 3$

12 $15x^2 - y^2 - 5x + 4y + 7 = 0$, $\quad y = 4x + 1$

13 $2x^2 + y^2 + 5x - 7y + 9 = 0$, $\quad 2x + y = 3$

14 $x^2 + 4y^2 + x + 2y - 58 = 0$, $\quad 5x - 2y + 2 = 0$

15 $2x^2 + 9y^2 - 10x - 36y - 1 = 0$, $\quad 3y = 5x - 2$

16 Find the coordinates of the points in which the straight line whose equation is $2x + y - 5 = 0$ intersects the circle with equation $x^2 + y^2 = 25$.

17 The straight line $x - y + 1 = 0$ cuts the ellipse $\dfrac{x^2}{8} + \dfrac{y^2}{18} = 1$ at A and B. Find the coordinates of A and B.

Miscellaneous Equations

EXERCISE 20b

Solve:

1 $4x^2 + y^2 - 6x - 3y = 16, \quad xy = 6$

2 $x^2 + y^2 + 3x - 5 = 0, \quad 3x^2 - xy - 2y^2 = 0$

3 $11x^2 - 10xy - 5y^2 = 19, \quad 2x^2 - xy - 2y^2 = 4$

4 $x^2 + y^2 = 5, \quad xy = -2$

5 $x^2 + y^2 = 13, \quad xy = 6$

6 $\dfrac{x}{y} + \dfrac{y}{x} = \dfrac{13}{6}, \quad xy = 6$

7 $2x^2 + 3y^2 = 35, \quad xy = 6$

8 $x + y = 2, \quad x^3 + y^3 = 26$

9 $xy = 3, \quad x^2 + x + y = 13$

10 The rectangular hyperbola $xy = 8$ intersects the circle $x^2 + y^2 = 20$. Find the coordinates of the points of intersection.

11 Find the coordinates of the points of intersection of the curves with equations $x^2 + y^2 = 20$ and $\dfrac{x^2}{4} + \dfrac{y^2}{36} = 1$.

Induction

21

EXERCISE 21

Prove the following results by induction.

1 $\displaystyle\sum_{r=1}^{n} r(r - 1) = \frac{n}{3}(n^2 - 1)$

2 $\displaystyle\sum_{r=1}^{n} r^2 = \frac{n}{6}(n + 1)(2n + 1)$

3 $\displaystyle\sum_{r=1}^{n} r^3 = \left[\frac{n(n + 1)}{2}\right]^2$

4 $1 \times 3 + 2 \times 4 + 3 \times 5 + \ldots n \times (n + 2) = \dfrac{n}{6}(n + 1)(2n + 7)$

5 $\displaystyle\sum_{r=1}^{n} \frac{1}{r(r+1)} = \frac{n}{n+1}$

6 $1 + 3 + 5 + 7 + \ldots + (2n-1) = n^2$

7 $a + (a+d) + (a+2d) + \ldots + [a + (n-1)d] = \dfrac{n}{2}[2a + (n-1)d]$

8 $a + ar + ar^2 + \ldots + ar^{n-1} = \dfrac{a(r^n - 1)}{r-1}$

9 $\displaystyle\sum_{1}^{n}(r+2) = \frac{n}{2}(n+5)$

10 Show by induction, or otherwise, that $3^{2n} - 1$ is divisible by 8 for all positive integer values of n.

Differentiation

USEFUL FACTS

$$\frac{d}{dx}(x^n) = nx^{n-1}$$

$$\frac{d}{dx}(\sin x) = \cos x$$

$$\frac{d}{dx}(\cos x) = -\sin x$$

$$\frac{d}{dx}(\tan x) = \sec^2 x$$

$$\frac{d}{dx}(\sec x) = \sec x \tan x$$

$$\frac{d}{dx}(\operatorname{cosec} x) = -\operatorname{cosec} x \cot x$$

$$\frac{d}{dx}(\cot x) = -\operatorname{cosec}^2 x$$

$$\frac{d}{dx}[\sin f(x)] = f'(x) \cos f(x)$$

with similar results for the other five trigonometric functions.

$$\frac{d}{dx}(\sin^{-1} x) = \frac{1}{\sqrt{1 - x^2}}$$

$$\frac{d}{dx}(\cos^{-1} x) = -\frac{1}{\sqrt{1 - x^2}}$$

$$\frac{d}{dx}(\tan^{-1} x) = \frac{1}{1 + x^2}$$

$$\frac{d}{dx}(a^x) = a^x \ln a$$

$$\frac{d}{dx}(e^{ax}) = ae^{ax}$$

$$\frac{d}{dx}(\ln f(x)) = \frac{f'(x)}{f(x)}$$

$$\frac{d}{dx}(uv) = u\frac{dv}{dx} + v\frac{du}{dx}$$

$$\frac{d}{dx}\left(\frac{u}{v}\right) = \frac{v\dfrac{du}{dx} - u\dfrac{dv}{dx}}{v^2}$$

$$\frac{d}{dx}(uvw) = uv\frac{dw}{dx} + vw\frac{du}{dx} + uv\frac{dw}{dx}$$

Simple Differentiation

EXERCISE 22a

Differentiate each function with respect to x.

1 $x^4 + 3x^2$ **2** $(x + 3)^5$ **3** $(2x + 4)^7$

4 $(x^2 - 3)^6$ **5** $\dfrac{1}{5x + 2}$ **6** $\dfrac{1}{(3 - 2x)^3}$

7 $x^2 + 5 + \dfrac{3}{x}$ **8** $\dfrac{x^3 - 4x}{x^2}$ **9** $\sqrt{4x + 2}$

10 $\sqrt{5 - x^2}$ **11** $\dfrac{3x - 4}{\sqrt{x}}$ **12** $6x\sqrt{x}$

Products and Quotients

EXERCISE 22b

Differentiate each function with respect to x.

1 $x^2(4x + 1)^3$ **2** $x(5 - x)^3$ **3** $\sqrt{x}\,(7x^2 + 2)^2$

4 $\dfrac{x}{1 + x}$ **5** $\dfrac{x}{(1 + x)^2}$ **6** $\dfrac{4x}{(1 - x)^2}$

7 $\dfrac{x^2}{1 + x^2}$ **8** $\dfrac{2x^2}{1 - x^2}$ **9** $\dfrac{x}{\sqrt{1 + x^2}}$

10 $\dfrac{3x}{\sqrt{1 - x}}$ **11** $\dfrac{\sqrt{x + 1}}{x}$ **12** $\sqrt{\dfrac{x - 1}{x + 1}}$

Trigonometric Functions

EXERCISE 22c

Differentiate each function with respect to x.

1 $\cos \dfrac{x}{2}$ **2** $\sin^2 2x$ **3** $\cos^2 3x$

4 $\tan 2x$ **5** $\sec \dfrac{x}{2}$ **6** $3 \operatorname{cosec} 3x$

7 $2 \cot \dfrac{x}{2}$ **8** $\sec^2 2x$ **9** $\tan^2 \dfrac{x}{2}$

10 $\sin x \cos 2x$ **11** $\dfrac{\sin x}{\cos \frac{x}{2}}$ **12** $\sin^2 x \cos x$

13 $\sin \left(\dfrac{\pi}{2} - \dfrac{x}{4} \right)$ **14** $\dfrac{\tan x}{\sec x}$ **15** $\dfrac{\cos^2 x}{\sin x}$

16 $\tan^3 x$ **17** $\sec x \operatorname{cosec} x$ **18** $\sqrt{\sin x}$

19 $\sqrt{\sec 2x}$ **20** $\dfrac{1}{\sqrt{\tan \frac{x}{2}}}$ **21** $\dfrac{\sin x}{\sqrt{\cos x}}$

22 $x \sec x$ **23** $\dfrac{\cos x}{\sqrt{x}}$ **24** $x^2 \tan x$

Logarithmic and Exponential Functions

EXERCISE 22d

Differentiate each function with respect to x.

1 $\ln 3x$ **2** $\ln x^4$ **3** $\ln \left(\dfrac{1}{x} \right)$

4 $\ln (4x + 1)$ **5** $\ln \sin x$ **6** $\ln \cos 2x$

7 $\ln \tan \dfrac{x}{2}$ **8** $x \ln x$ **9** $\ln \sec \dfrac{x}{2}$

10 e^{3x} **11** e^{2x^2} **12** e^{-4x}

13 $e^x \sin x$ **14** $\dfrac{1}{e^{3x}}$ **15** $e^{-x} \cos x$

16 $\dfrac{\tan x}{e^x}$ **17** $\dfrac{e^{2x} - 1}{e^x}$ **18** xa^x

19 $\ln \left(\dfrac{3x + 1}{x + 1} \right)$ **20** $\ln \left(\dfrac{1}{4x + 3} \right)$ **21** $\ln \left(\dfrac{\sin x}{\cos 2x} \right)$

22 $\ln (1 + x) \sin x$ **23** $\ln (1 - x^2) \tan x$ **24** $\ln \sec x \tan x$

Implicit Functions

If $x^2 + y^3 = 1$,

$$2x + 3y^2 \frac{dy}{dx} = 0$$

i.e.

$$\frac{dy}{dx} = -\frac{2x}{3y^2}$$

EXERCISE 22e

1 If $x^2 + y^2 = a^2$ show that $\dfrac{dy}{dx} = -\dfrac{x}{y}$.

2 Find $\dfrac{dy}{dx}$ at the point $(-4, 3)$ on the circle $x^2 + y^2 = 25$.

3 Find $\dfrac{dy}{dx}$ when:

a) $x^2 y^3 = 8$,

b) $x^2 + y^3 = 8$,

c) $\dfrac{x^2}{y^3} = 8$.

4 Find $\dfrac{dy}{dx}$ and $\dfrac{d^2y}{dx^2}$ at the point $(2, -1)$ on the circle $x^2 + y^2 = 5$.

5 If $3(x - y)^2 + xy = 15$ find $\dfrac{dy}{dx}$ at the point $(3, 1)$.

Parameters

If $x = f(t)$ and $y = g(t)$, in general:

$$\frac{dy}{dx} = \frac{dy}{dt} \times \frac{dt}{dx} \text{ where } \frac{dt}{dx} = 1 \bigg/ \frac{dx}{dt}$$

EXERCISE 22f

Find $\dfrac{dy}{dx}$ in terms of the parameter when:

1 $x = t^2, \quad y = t^3$
    2 $x = ct, \quad y = \dfrac{c}{t}$

3 $x = \dfrac{a}{t^2}, \quad y = -\dfrac{2a}{t}$
    4 $x = a \sin \theta, \quad y = a \cos \theta$

5 $x = a \sec \theta, \quad y = b \tan \theta$
    6 $x = a\dfrac{1 + t^2}{1 - t^2}, \quad y = b\dfrac{2t}{1 - t^2}$

7 Find, in terms of t, $\dfrac{dy}{dx}$ and $\dfrac{d^2y}{dx^2}$, when $x = t + t^2$ and $y = t^2 + t^3$.

8 If $x = 4t$ and $y = 3t - 5t^2$ find $\dfrac{dy}{dx}$ in terms of t. For what value of t is $\dfrac{dy}{dx} = 0$?

Show that $\dfrac{d^2y}{dx^2} = -\dfrac{5}{8}$ when $\dfrac{dy}{dx} = 0$.

9 A curve is defined by the parameter equations $x = t + \dfrac{1}{t}$; $y = t - \dfrac{1}{t}$. Find $\dfrac{dy}{dx}$ and $\dfrac{d^2y}{dx^2}$, each in terms of t.

Rate of Change

EXERCISE 23

1 If $x^2 + y^2 = 25$ find the value of $\dfrac{dy}{dx}$ when $x = -3$ and $y = 4$. Can you attach a meaning to this value? Illustrate your answer with a diagram.

2 The side of a cube is increasing at the rate of 3 cm/second. Find the rate of increase of the volume when the length of a side is 5 cm.

3 A rectangle is three times as long as it is wide. Find the rate of increase of the perimeter when the rectangle is 30 cm long if its area is growing at the rate of 10 cm²/second.

4 The area of a circular patch of oil on the surface of a pool starts from zero and increases at the rate of 10 cm²/second. Find the rate of increase of the radius:
a) after 5 seconds,
b) when the radius of the circular patch is 5 cm.

5 A point moves in a straight line so that its displacement, s metres, from a fixed point O, at time t seconds, is given by $s = t^3 - 6t^2 + 5t$.
a) Find its velocity and acceleration when: i) $t = 0$, ii) $t = 2$, iii) $t = 4$.
b) On how many occasions, and at what times does the particle pass through O after the start?

6 A right circular cylinder has a constant volume of 300 cm³. Its radius increases at the rate of 1 cm/second. Find the rate of change of the height when the radius is 5 cm. What is the significance of the minus sign?

7 A gas law can be given in the form $pv^{1.4} = c$ where c is a constant. If p increases by 1% find the corresponding change in v, stating clearly whether there is an increase or a decrease.

8 The constant volume of a right circular cone is 308 cm³. Find the value of its height when the radius is 7 cm ($\pi = \frac{22}{7}$). When the radius is 7 cm the radius is decreasing at the rate of 5 mm/second. Find the rate of change of the height at the same time.

9 Wheat is dropped from an elevator shaft on to the ground at a steady rate of 20 m³/minute. It forms a conical pile whose height remains equal to the radius of its base. At what rate is the height increasing:
a) when the height is 10 metres,
b) after 10 minutes?

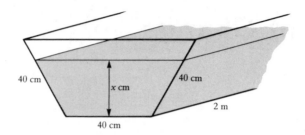

A horse trough is made using three similar planks each 40 cm wide and 2 m long. They are put together so that the cross-section of the trough is a trapezium, the sloping sides making angles of 30° with the vertical.

Show that when the depth of water is x cm the volume of water in the trough is

$$200x \left(40 + \frac{x}{\sqrt{3}} \right) cm^3.$$

A horse approaches the trough when it is full and consumes water at a rate of 4.8 litres per minute. Find the rate, in cm/min, at which the water level is falling when the depth of water in the trough is 30 cm.

Stationary Values— Curve Sketching

24

EXERCISE 24

For each of the functions $f(x)$ given in questions 1 to 6 find:

a) where the graphs cross the x and y axes,
b) the coordinates of any stationary points.

Use this information to sketch the graph of $f(x)$.

1 $f(x) = x^2 - 5x + 4$ **2** $f(x) = 18 - 3x - x^2$

3 $f(x) = x^3 - 5x^2 + 4x$ **4** $f(x) = 12x - x^2 - x^3$

5 $f(x) = (x-1)(x-2)(x-4)$ **6** $f(x) = (x-2)^2 - 9$

7 Sketch the graph of $f(x) = x^2 - 5x + 6$ and use it to sketch, on the same axes, the graph of $\dfrac{1}{f(x)}$.

8 The curve whose equation is $y = x^3 - ax^2 + bx + c$ passes through the point $(2, 7)$, has a local maximum when $x = 1$ and a point of inflexion when $x = 3$. Find the values of a, b and c. Find the coordinates of the minimum point on this curve. Sketch the curve.

9 The graph of $y = ax^3 + bx + c$ has a point of inflexion when $y = 3$, a local maximum when $x = 1$ and passes through the point $(2, 1)$. Find the values of a, b and c and give the value of x at which the graph has a local minimum. Sketch the graph.

10

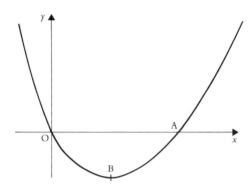

The sketch shows the graph of $y = f(x)$. The graph crosses the x-axis at the origin and at the point A $(a, 0)$. The minimum point on the curve is B (h, k). Write down:

a) the coordinates of the points where the graph of $y = f(x - a)$ crosses the x-axis,

b) the coordinates of the minimum point on the curve $y = f(x + a)$,

c) the coordinates of the maximum point on the curve $y = -f(x - a)$.

11

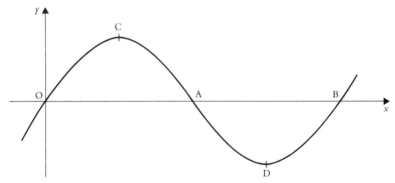

The diagram shows the graph of $f(x) = x(x - a)(x - 2a)$. C is a maximum point on the curve and has coordinates $\left(\dfrac{a}{2}, b\right)$ and D is a minimum point on the curve and has coordinates $\left(\dfrac{3a}{2}, -b\right)$.

a) Write down the coordinates of A and B, two of the points where the graph crosses the x-axis.

b) Sketch the graph of $y = f(x + a)$. Write down the coordinates of the points at which this graph crosses the x-axis, and the coordinates of any maximum or minimum points.

c) Sketch the graph of $y = f(2x)$ showing clearly the coordinates of the maximum and minimum points and the coordinates of the points where the graph crosses the x-axis.

12 Show that the expression $y = \dfrac{4x}{(4x + 1)(x - 1)}$ does not have any real values

of x for which $\dfrac{dy}{dx} = 0$. Sketch the curve showing clearly any horizontal or vertical asymptotes.

13 The function f is defined by $f(x) = \dfrac{4}{x - 1} - \dfrac{1}{x + 2}$.

Show that $f'(x) = -\dfrac{3(x + 1)(x + 5)}{(x - 1)^2(x + 2)^2}$.

Find the stationary values of f and sketch its graph.

14 Sketch, on separate diagrams, the graphs of $y = \dfrac{x^2}{x^2 - 4}$ and $y = x^2(x^2 - 4)$.

Show clearly any maximum or minimum points and any horizontal or vertical asymptotes.

15 The function f is defined by $f(x) = \dfrac{x^2}{x + 1}$.

Show that f cannot take any value between -4 and 0. Sketch the graph of f showing any stationary values.

Use your sketch of f to draw on the same diagram the sketch of g where g is defined by $g(x) = \dfrac{1 + x}{x^2}$.

16 The function f is defined by $f(x) = \dfrac{x^2}{1 - x^2}$.

Show that f cannot take any value between -1 and 0. Sketch the graph of f. Is the graph symmetrical about the y-axis? Give a reason for your answer.

A second function g is defined by $g(x) = \dfrac{1 - x^2}{x^2}$.

Use your sketch for f to draw a sketch for g. You can show both sketches on the same axes.

17 a) Sketch the curve whose equation is $y = \dfrac{x}{1 + x^2}$. Find the equation of the tangent and the equation of the normal at the point where $x = 1$.

b) Use your answer to part (a) to sketch the curve whose equation is

$y = \left| \dfrac{x}{1 + x^2} \right|$. What is the equation of the tangent to this curve at the

point $(-1, \frac{1}{2})$?

18 Sketch the curve with equation $y = \dfrac{x}{x-1}$. Hence sketch the curve with

equation $y = \left| \dfrac{x}{x-1} \right|$.

19 Sketch the curve $y = \tan x$ for $-\pi \leqslant x \leqslant \pi$. On the same axes sketch the graph of $y = \cot x$ showing clearly where the two graphs cross. Use these graphs to sketch on separate axes, for $-\pi \leqslant x \leqslant \pi$, the curve $y = \tan x + \cot x$.

20 a) Show that, if x is real,
$$-2 \leqslant x + \frac{1}{x} \leqslant 2$$

Hence sketch the graph of $y = f(x)$ where $f(x) = x + \dfrac{1}{x}$. Show clearly any maximum or minimum points, together with any vertical or horizontal asymptotes.

b) Use your graph of $y = f(x)$ to sketch the graph of

i) $y = \dfrac{1}{f(x)}$ ii) $y = f(x + 1)$ iii) $y = f(2x)$.

The following graphs involve moduli. Sketch the graph of:

21 $y = 3x + 1$ and use it to sketch the graph of $y = |3x + 1|$

22 $y = 1 - 2x$ and use it to sketch the graph of $y = |1 - 2x|$

23 $y = |x^2 - 2|$

24 $y = |\sin x|$ $0 \leqslant x \leqslant 2\pi$

25 $y = |\sin x + \cos x|$ $0 \leqslant x \leqslant 2\pi$

26 $y = |6 - x - x^2|$

27 $y = 2 + |x + 2|$

28 $y = |x(x - 1)(x - 2)|$

Problems Involving Maxima and Minima   25

EXERCISE 25

1 Prove that if the sum of the perimeters of two squares is 100 cm, the sum of the areas is least when the squares are of equal size.

2 The sum of the length and girth of a rectangular block of metal with square ends is not to exceed 18 cm. Find the largest volume that the block can have.

3

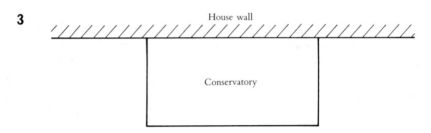

Leroy wishes to attach a conservatory with a ground area of 18 m^2 to the back of his bungalow. Find its dimensions so that the length of its outside wall is a minimum.

4 A closed rectangular tank, with square ends, is to have a capacity of 8 m^3. Find its dimensions if its total external surface area is a minimum.

5

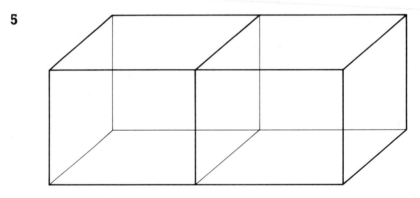

The sketch shows a frame made from 180 cm of stiff wire. It is in the shape of a cuboid with a band around the middle, and is to be used to make a cage for mice. If the cage has square ends of side x cm and is l cm long find an expression for l in terms of x. Hence show that the volume enclosed by the cage is V cm^3 where $V = x^2(45 - 3x)$.

Find the value of x for which V is a maximum and the corresponding value of V.

6 A right circular cylinder is inscribed in a right circular cone of radius r and height h, such that one circular end of the cylinder rests on the base of the cone. Show that the maximum possible volume of the cylinder is four-ninths the volume of the cone.

7 A church window of area $43\frac{3}{4}\,\text{m}^2$ is in the form of a rectangle surmounted by a semicircle. Show that the perimeter of the window is a minimum when the width of the window is equal to its height. (Take $\pi = \frac{22}{7}$.)

8 A cylinder is such that the sum of its diameter and height is 20 cm. Express the volume ($V\,\text{cm}^3$) in terms of the radius of its base (r cm). What is the greatest possible value for the volume of the cylinder?

9 A 6 m length of wire is cut into two pieces. The first piece is formed into a square while the other piece is bent to give a circle. Find the dimensions of each if the sum of the areas they enclose is a minimum.

10 The equal sides of an isosceles triangle, with a constant perimeter of $2a$, are each of length x. Find an expression, in terms of a and x, for:
a) the base of the triangle,
b) the perpendicular height of the triangle.

Hence write down an expression for the area of the triangle and prove that its area is a maximum when $x = \frac{2}{3}a$.

11 Find the volume of the largest right circular cylinder that can be cut from a sphere of radius a.

12

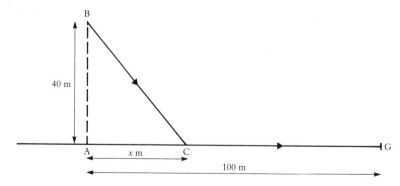

A farmer wishes to build a battery house for his hens at B, 40 m from the nearest point, A, on the lane to the main gate G. It costs £25 per metre to lay a driveway across the land from B to the lane AG, and £15 per metre to lay a surface on the existing lane. Find an expression for the total cost £C if he lays the driveway from B to C (where AC $= x$ m), and on to G. Hence find the value of x that gives the cheapest driveway. How much will it cost?

53

13 A cylindrical mug has a constant internal surface area A and the radius of its base is r. Show that the capacity of the mug, C, is given by $C = \dfrac{r}{2}(A - \pi r^2)$ cubic units. Prove that the capacity is a maximum when the depth of the mug is equal to its radius.

The Circle **26**

The equation:

$$x^2 + y^2 = a^2$$

represents a circle with centre at the origin, radius a.

The equation:

$$x^2 + y^2 + 2gx + 2fy + c = 0$$

represents a circle with centre $(-g, -f)$, radius $\sqrt{g^2 + f^2 - c}$.

The equation of the tangent to the circle $x^2 + y^2 + 2gx + 2fy + c = 0$ at the point (x_1, y_1) is:

$$xx_1 + yy_1 + g(x + x_1) + f(y + y_1) + c = 0$$

The equation of the normal to the same circle at the point (x_1, y_1) is:

$$\frac{y - y_1}{x - x_1} = \frac{y_1 + f}{x_1 + g}$$

EXERCISE 26

1 Find the equation of the circle:

  a) with the origin as the centre and radius 3,
  b) with the point $(2, 3)$ as centre and radius 2,
  c) with the point $(4, -2)$ as centre and radius 4,
  d) with the point $(-3, -2)$ as centre and radius 5,
  e) with the point $(\frac{1}{2}, -3)$ as centre and radius $\frac{3}{2}$.

2 Find the centre and radius of the circle with equation:

  a) $x^2 + y^2 = 4$ b) $(x - 2)^2 + (y - 3)^2 = 25$
  c) $(x + 3)^2 + (y - 4)^2 = 1$ d) $(x + 1)^2 + (y + 5)^2 = 16$
  e) $4x^2 + 4y^2 = 25$ f) $(2x - 1)^2 + (2y + 1)^2 = 16$

3 Find the centre and radius of the circle with equation:

  a) $x^2 + y^2 + 4x - 6y - 3 = 0$ b) $x^2 + y^2 - 5x + 3y + 4 = 0$
  c) $9x^2 + 9y^2 = 25$ d) $5x^2 + 5y^2 - 10x + 20y - 11 = 0$

4 Determine whether or not the given point is inside, on, or outside the given circle.

a) $(-2, 1)$, $\quad x^2 + y^2 = 4$
b) $(-3, -4)$, $\quad x^2 + y^2 = 25$
c) $(7, 3)$, $\quad (x - 3)^2 + (y - 1)^2 = 25$
d) $(4, 3)$, $\quad x^2 + y^2 - 4x + 2y - 11 = 0$

5 The circle $x^2 + y^2 + 2gx + 2fy + c = 0$ passes through the points $(-2, 1)$, $(2, -3)$ and $(6, 1)$. Find the values of g, f and c.

6 Find the equation of the circle which has as diameter the join of the points $(-2, -2)$ and $(3, 4)$.

7 Find the equation of the circle passing through the points $(3, 0)$, $(3, 2)$ and $(-1, 2)$. Write down the centre and radius of this circle.

8 Prove that the points $(6, 2)$, $(2, 4)$, $(-3, -1)$ and $(6, -4)$ lie on a circle. Write down the equation of this circle, the coordinates of its centre and its radius.

9 Write down the equation of the tangent to the given circle at the given point.

a) $x^2 + y^2 = 25$, $\quad (4, 3)$
b) $x^2 + y^2 - 2x - 2y - 3 = 0$, $\quad (3, 2)$
c) $x^2 + y^2 + 4x + 2y - 20 = 0$, $\quad (-6, 2)$
d) $x^2 + y^2 - 6x - 4y = 0$, $\quad (6, 4)$

10 Find the points at which the straight line with equation $2x + 3y = 6$ cuts the circle with equation $x^2 + y^2 - 3x - 2y = 0$.

11 The straight line with equation $x + y = 4$ cuts the circle $x^2 + y^2 + 4x - 4y - 8 = 0$ at two points A and B. Find the coordinates of A and B, and hence find the length of the chord AB.

12 Show that the circle $x^2 + y^2 - 2x - 2y - 3 = 0$ trisects the line joining the points $(-4, 1)$ and $(5, 4)$.

13 The line $3x + y - 4 = 0$ intersects the circle $x^2 + y^2 - 2x - 2y - 8 = 0$ at A and B. Show that the chord AB subtends a right angle at the point C $(4, 2)$. Does C lie on the circle?

14 The centre of the circle $x^2 + y^2 + 2gx + 2fy + c = 0$ lies on the line $2x + y = 4$, and the circle passes through the points $(-1, 0)$ and $(-1, 4)$.

a) Find i) the coordinates of the centre of the circle
ii) its radius.
b) Show that the chord joining the two given points subtends a right angle at the centre of the circle.

15 Find the equations of the circles passing through the points $(0, 1)$ and $(8, 1)$ that have a radius of 5.

16 Find the equation of the circle that passes through the point $(3, -5)$ and touches the line $3x + 2y = 12$ at the point $(4, 0)$.

17 Show that each of the given lines is a tangent to the given circle.
a) $4x - 3y - 24 = 0$, $x^2 + y^2 = 25$
b) $4x - y - 3 = 0$, $x^2 + y^2 + 6x - 4y - 4 = 0$
c) $7x - 2y + 16 = 0$, $x^2 + y^2 - 3x - 11 = 0$
d) $4x + y - 11 = 0$, $x^2 + y^2 + 2x + 4y - 12 = 0$

18 Find the equations of the tangents from the point $(5, 5)$ to the circle $x^2 + y^2 = 5$. Show that the angle between these tangents is $\tan^{-1} \frac{3}{4}$.

19 Find the equations of the tangents from the point $(2, 3)$ to the circle $x^2 + y^2 + x - y - 2 = 0$. What is the size of the acute angle between these two tangents?

20 Show that the straight line with equation $x + 3y + 5 = 0$ is a tangent to the circle $x^2 + y^2 + 2x - 4y - 5 = 0$.

Show that this tangent passes through the point $P(4, -3)$ and find the equation of the other tangent from P to the circle.

Find the acute angle between the two tangents and deduce that the angle subtended at the centre of the circle by the straight line joining the points of contact of the tangents with the circle is about 127°.

21 Two circles are denoted by C_1 and C_2. The equation of C_1 is $x^2 + y^2 + 2x + 4y - 20 = 0$ and the equation of C_2 is $x^2 + y^2 - 10x - 5y + 25 = 0$.
a) Find the coordinates of the centre and the radius for C_1 and C_2.
b) Find the distance between the centres of the two circles.
c) Deduce that C_1 and C_2 touch and write down the coordinates of the point of contact.

22 The equations of two circles S_1 and S_2 are:

S_1: $\qquad\qquad\qquad x^2 + y^2 - 4x - 2y - 20 = 0$

S_2: $\qquad\qquad\qquad x^2 + y^2 - 18x - 4y + 60 = 0$

Find the coordinates of A, the point of intersection of these two circles in the first quadrant.
Find the equation of the tangent to S_1 at A and the equation of the tangent to S_2 at A.
Write down the equations of the normals to S_1 and S_2 at A.

23 The circle $x^2 + y^2 = a^2$ cuts the x-axis at A and C and the y-axis at B and D, A and B being on the positive axes. Write down the coordinates of A, B, C and D.

A fifth point P with coordinates $(a \cos \theta, a \sin \theta)$ varies as θ varies but does not lie on either axis. Show that the equation of PD is $x(1 + \sin \theta) - y \cos \theta - a \cos \theta = 0$ and find the equations of AB and CP.

CP crosses the y-axis at M and PD and AB intersect at N. Show that the y-coordinate of both M and N is $\dfrac{a \sin \theta}{1 + \cos \theta}$. Hence write down the equation of MN.

24 Find the centres and radii of the circles:

C_1: $\qquad\qquad x^2 + y^2 - 4x = 0$

C_2: $\qquad\qquad x^2 + y^2 - 4y = 0$

In addition to the origin the straight line $y = 2x$ cuts C_1 at A and C_2 at B. Show the circles and straight line in a sketch indicating also the position of C, the second point of intersection of the two circles.

Find the coordinates of A, B and C. Hence show that AC and BC are perpendicular.

Further Coordinate Geometry

27

The Parabola

The equation $y^2 = 4ax$ represents a parabola that passes through the origin and is symmetrical about the x-axis.

Sometimes both x and y are given as functions of another variable. This variable is called a parameter. The point $(at^2, 2at)$ lies on the parabola $y^2 = 4ax$ for any value of the parameter t.

The equation of the tangent to the parabola $y^2 = 4ax$ at the point t is $y = \dfrac{x}{t} + at$ and the equation of the normal at the same point is $y + tx = 2at + at^3$.

The line $y = mx + \dfrac{a}{m}$ touches the parabola for all values of m.

The normal at the point t meets the parabola again at the point $-\dfrac{2}{t} - t$.

EXERCISE 27a

1 Find the equation of: a) the tangent, b) the normal, to the parabola $y^2 = 8x$ at the point $(2, 4)$.

2 Find the equation of: a) the tangent, b) the normal, to the parabola $y^2 = 4ax$ at the point $(a, 2a)$.

3 $P(ap^2, 2ap)$ and $Q(aq^2, 2aq)$ are two points on the parabola $y^2 = 4ax$. Find the equation of the chord PQ.

If the chord PQ subtends a right angle at the point $R(ar^2, 2ar)$ show that $(p + r)(q + r) = -4$.

4 Show that the chord joining the points $P(ap^2, 2ap)$ and $Q\left(\dfrac{a}{p^2}, -\dfrac{2a}{p}\right)$ passes through the point $(a, 0)$.

Show that the equation of the locus of the midpoint of PQ is the parabola $y^2 = 2a(x - a)$.

5 A variable chord of the parabola $y^2 = 4ax$ passes through the fixed point $(3a, 0)$. Show that the locus of the midpoint of the chord is the parabola $y^2 = 2a(x - 3a)$. Find the equation of the locus of the point of intersection of the tangents drawn at the ends of this variable chord.

6 $P(ap^2, 2ap)$ and $Q(aq^2, 2aq)$ are two points on the parabola $y^2 = 4ax$.

Prove that:

a) the equation of the chord PQ is $y(p + q) = 2x + 2apq$,

b) the equation of the tangent at P is $y = \dfrac{x}{p} + ap$,

c) the tangents at P and Q intersect at $[apq, a(p + q)]$.

7 Show that the tangents to the parabola $y^2 = 4ax$ at the points $P(ap^2, 2ap)$ and $Q(aq^2, 2aq)$ intersect at the point $T[apq, a(p + q)]$.

If PT and QT are perpendicular show that T lies on the line $x = -a$.

8 Find the equation of the tangent to curve $y^2 = 4ax$ at the point $P(ap^2, 2ap)$ and the equation of the normal to the curve at the point $Q(aq^2, 2aq)$. Find the relationship between p and q if these lines are:

a) parallel, b) perpendicular.

9 Find the equation of the tangent to the parabola $y^2 = 4x$ at the point P (4, 4). The straight line joining the point P to the point A (2, 0) is produced to meet the parabola again at Q.

Find the coordinates of Q and find the equation of the tangent to the parabola at Q.

Show that the tangents at P and Q intersect at T $(-2, 1)$ and find the angle PTQ giving your answer correct to the nearest tenth of a degree.

10 Show that the line $y = 2x + 1$ touches the parabola $y^2 = 8x$ and find the coordinates of the point of contact P.

Find the equation of the tangent to this parabola at the point Q $(8, -8)$ and the equation of the chord PQ.

The tangents at P and Q intersect at T and the chord PQ crosses the x-axis at R. Find the coordinates of T and R and show that the midpoint of TR lies on the y-axis.

11 Find the equation of the tangent to the curve $y^2 = 4ax$ at the point P $(at^2, 2at)$.

The straight line through the origin parallel to this tangent meets the curve again at Q. Find the coordinates of R, the midpoint of OQ and show that PR is parallel to the x-axis.

12 P $(ap^2, 2ap)$ and Q $(aq^2, 2aq)$ are two points on the parabola $y^2 = 4ax$.

Prove that: a) the equation of the normal at P is $y + px = 2ap + ap^3$,
               b) the normals at P and Q intersect at the point
                $[a(p^2 + pq + q^2 + 2), -apq(p + q)]$.

What is the condition that the normals at P and Q are perpendicular?

If the normals at P and Q are perpendicular and intersect at N show that N lies on the curve whose equation is $y^2 = a(x - 3a)$.

13 Show that the equation of the normal to the parabola $y^2 = 4ax$ at the point P $(ap^2, 2ap)$ has equation $y + px = 2ap + ap^3$.

Q is the point $\left(\dfrac{a}{p^2}, -\dfrac{2a}{p}\right)$. Show that the chord PQ passes through the point $(a, 0)$.

Find the coordinates of M, the midpoint of PQ.

The line through M parallel to the axis of the parabola meets the normal through P at N. Show that the locus of N is the parabola $y^2 = a(x - 3a)$.

Other Curves

EXERCISE 27b

1 Find the equation of: a) the tangent, b) the normal, to the curve $y = x^3 - 3x - 1$ at the point $(2, 1)$.

2 A curve has equation $y = 2x^3 - 3x^2 - 12x + 5$. Find the values of x for which $\dfrac{dy}{dx} = 0$. Hence find the equations of the tangents to this curve that are parallel to the x-axis.

3 Find the points of intersection of the curve $y = x^3 - x^2 - 6x$ with the x-axis. Find the equation of the tangent to the curve at each of these points. Show that the two tangents with positive gradients intersect at the point $(13, 150)$.

4 Find the gradient of the tangent to the curve $y = 2 - x^2 - x^3$ at the point $P(-2, 6)$. Hence find the equation of the tangent and the equation of the normal at the point P. The tangent meets the y-axis at A and the normal meets the y-axis at B. Show that the length of AB is $16\frac{1}{4}$ units.

5 Find the equation of the tangent and of the normal to the curve with equation $\dfrac{x^2}{8} + \dfrac{y^2}{18} = 1$ at the point $(2, -3)$.

6 Prove that the straight line $y = 2x - 5$ is a tangent to the curve $9x^2 + 4y^2 = 36$.

7 Show that the straight line $y = mx + c$ is a tangent to the ellipse $\dfrac{x^2}{a^2} + \dfrac{y^2}{b^2} = 1$ if $c^2 = a^2m^2 + b^2$.

8 Find the equation of the tangent to the curve $x = a \cos \theta$, $y = b \sin \theta$ at the point $\theta = \frac{\pi}{4}$.

9 $P(a \cos \theta, b \sin \theta)$ and $Q(a \cos \phi, b \sin \phi)$ are two points on the ellipse $\dfrac{x^2}{a^2} + \dfrac{y^2}{b^2} = 1$. Find the equations of the tangents at P and Q and show that they intersect at the point $T\left(\dfrac{a \cos \frac{1}{2}(\theta + \phi)}{\cos \frac{1}{2}(\theta - \phi)}, \dfrac{b \sin \frac{1}{2}(\theta + \phi)}{\cos \frac{1}{2}(\theta - \phi)}\right)$.

If $\theta = \frac{\pi}{3}$ and $\phi = \frac{\pi}{6}$ show that T lies on the line $y = \dfrac{b}{a} x$.

10 Find the equation of the normal to the curve $x^2 + 4y^2 = 4$ at the point $P\left(\sqrt{2}, \dfrac{\sqrt{2}}{2}\right)$. This normal crosses the x-axis at N. If M is the foot of the perpendicular from P to the x-axis show that $ON.OM = \frac{3}{2}$.

11 Show that the point $P\left(\dfrac{a(1-t^2)}{1+t^2}, \dfrac{2bt}{1+t^2}\right)$ lies on the curve $\dfrac{x^2}{a^2}+\dfrac{y^2}{b^2}=1$, and find the equation of the tangent to the curve at P in the form $y = mx + c$.

12 Prove that the line $ax + by + c = 0$ will touch the curve $9x^2 - 4y^2 = 36$ if $4a^2 - 9b^2 = c^2$.

13 Show that the equation of the tangent to the curve $b^2x^2 - a^2y^2 = a^2b^2$ at the point $P\,(a\sec\theta, b\tan\theta)$ is $bx\sec\theta - ay\tan\theta - ab = 0$. This tangent meets the x-axis at T. N is the foot of the perpendicular from P to the x-axis. Show that $ON.OT^2 = a^2$.

14 $P\left(ct, \dfrac{c}{t}\right)$ and $Q\left(ct_1, \dfrac{c}{t_1}\right)$ are two points on the curve $xy = c^2$. The tangents at P and Q intersect at T. Find the coordinates of T and show that if $tt_1 = K$ the locus of T is a straight line passing through the origin. Find the equation of this locus.

Functions

28

The set of input values for which a function is defined is called the domain of that function. The corresponding set of output values is called the range of the function or the image set.

If a function f maps the domain of f on to the image set of f and if the reverse mapping of the image set of f to the domain of f is one-one, then the mapping is called the inverse function. It is denoted by f^{-1}.

The graph of $y = f^{-1}(x)$ will be a reflection of the graph of $y = f(x)$ in the line $y = x$.

$g(f(x))$ or $gf(x)$ or $g \circ f$ is called a composite function.

For example if $f(x) = x^2$ and $g(x) = 3x - 2$,

$$gf(x) = g(x^2) = 3x^2 - 2$$

and $$fg(x) = f(3x - 2) = (3x - 2)^2 = 9x^2 - 12x + 4$$

Functions that are symmetrical about the vertical axis are called even functions.

A function with the property that $f(-a) = -f(a)$ for every member a of the domain is called an odd function.

EXERCISE 28

1 Write down the image set of the set $\{-2, -1, 0, 1, 2\}$ under the mapping $x \rightarrow x^2$.

2 Write down the image set of the set $\{-3, -1, 0, 1, 2, 4\}$ under the mapping $x \rightarrow 4 - 3x$.

3 The function f maps x on to $f(x)$ where $f(x) = 7 - 2x$. If A is the set $\{1, 2, 4\}$ on to what set is A mapped by f?. Find the value of x if f maps x on to 1.

4 The domain of the function $f(x) = 4x - 7$ is $\{0, 1, 2, 3, 4, 5\}$. Find its range.

5 The domain of the function $g(x) = x^2 + 4$ is $(-\infty, \infty)$. Find its range.

6 The domain of the function $h(x) = 4 - x^2$ is $\mathbb{R}$. Find its range.

7 The domain of the function $f(x) = +\sqrt{16 - x^2}$ is a subset of $\mathbb{R}$. Write down the largest possible set that is a suitable domain. What is the corresponding range?

8 The function f maps x on to $f(x)$ where $f(x) = kx - 3$. Find k if f maps 3 on to 3.

9 The function f maps x on to $f(x)$ where $f(x) = 4 - kx$. Find k if f maps 4 on to 8.

10 The function f maps x on to $f(x)$ where $f(x) = 3 + 4x$ and the function g maps x on to $g(x)$ where $g(x) = 3x^2$.

Find:

a) $f(4)$ b) x if $f(x) = 23$ c) $g(-3)$ d) $f[g(-3)]$
e) $g[f(4)]$ f) $f[g(x)]$ g) $g[f(x)]$

11 The function f maps x on to $f(x)$ where $f(x) = 8 + 5x - 2x^2$ and the function g maps x on to $g(x)$ where $g(x) = 5x - 3$.

Find:

a) $f(4)$ b) $g[f(4)]$ c) $g(2)$ d) $f[g(2)]$ e) $g[f(x)]$
f) $f[g(x)]$

12 The function f maps x on to $f(x)$ where $f(x) = \dfrac{8}{3x - 2}$. If the domain of f is $\{0, 2, 4\}$ find its range.

13 The functions f, g and h are defined by $f(x) = x^3 - 2x^2$, $g(x) = 4 - x^2$, $h(x) = \sin x$. State which function is: a) an even function, b) an odd function, c) neither.

62

14 State whether each of the following functions is an even function, an odd function or neither.

a) $f(x) = \dfrac{4}{x}$ b) $g(x) = x^2 + 2$ c) $h(x) = -2x^3$

15 The functions f and g are defined by $f(x) = 5x + 3$ and $g(x) = x + 5$. Find $fg(x)$ and $(fg)^{-1}(x)$.

Show that if $fg(a) = b$ then $(fg)^{-1}(b) = a$.

16 Find the inverses of the functions f and g defined by $f(x) = \dfrac{1}{x-5}$ $(x \neq 5)$

and $g(x) = \dfrac{1}{5x + 4}$ $(x \neq -\tfrac{4}{5})$.

17 The function $f: x \to x^2$ is defined for the domain $[0, \infty)$. Show that f has an inverse f^{-1} and state the domain and range of f^{-1}. Show the graphs of f and f^{-1} in the same diagram.

18 The functions f and g are defined by $f(x) = +\sqrt{1 - 9x^2}$ $(-\tfrac{1}{3} \leqslant x \leqslant 0)$ and $g(x) = e^{\frac{x}{2}}$ $(-\infty < x < \infty)$.

Find expressions for f^{-1} and g^{-1}. Give the domain and range of each.

19 The domain of the function $f(x) = +\sqrt{9 - x^2}$ is $[-3, 3)$. What is the range of f? Does f have an inverse? If not suggest how the domain can be restricted so that it does. Using the restricted domain for f state the domain and range of f^{-1}.

20 The functions f, g and h are defined by

$$f(x) = \dfrac{4x}{x - 2}, \; g(x) = \ln(x + 2), \; h(x) = +\sqrt{25 - x^2}.$$

a) State the maximum domain for each function.
b) Find the composite function $g \circ f$ and give its maximum domain.
c) Find the inverse functions f^{-1} and g^{-1}. Give the maximum domain of each.
d) Explain why h^{-1} does not exist unless the maximum domain is restricted. Give a domain of h for which h^{-1} exists.

21 Sketch the graph of the function f defined by $f(x) = \dfrac{x}{1 + x^2}$ where x is real.

A function g, which is one–one, is defined by restricting the domain of f to $[-k, k)$. What is the largest possible value for k? Find g^{-1}. Write down the domain and range of g^{-1}.

Integration

29

USEFUL FACTS

$$\int x^n \, dx = \frac{x^{n+1}}{n+1} \quad (n \neq -1)$$

$$\int \sin x \, dx = -\cos x$$

$$\int \cos x \, dx = \sin x$$

$$\int \sec^2 x \, dx = \tan x$$

$$\int \operatorname{cosec}^2 x \, dx = -\cot x$$

$$\int \sec x \tan x \, dx = \sec x$$

$$\int \operatorname{cosec} x \cot x \, dx = -\operatorname{cosec} x$$

$$\int \tan x \, dx = -\ln|\cos x| = \ln|\sec x|$$

$$\int \cot x \, dx = \ln|\sin x|$$

$$\int \frac{dx}{\sqrt{1-x^2}} = \sin^{-1} x$$

$$\int \frac{dx}{1+x^2} = \tan^{-1} x$$

$$\int \frac{1}{x} \, dx = \ln|x|$$

$$\int \frac{1}{ax+b} \, dx = \frac{1}{a} \ln|ax+b|$$

$$\int \frac{f'(x)}{f(x)} \, dx = \ln|f(x)|$$

$$\int u \frac{dv}{dx} \, dx = uv - \int v \frac{du}{dx} \, dx$$

$$\int (ax+b)^n \, dx = \frac{(ax+b)^{n+1}}{a(n+1)} \quad (n \neq -1)$$

$$\int e^{ax+b} \, dx = \frac{1}{a} e^{ax+b}$$

$$\int \sin(ax + b)\, dx = -\frac{1}{a} \cos(ax + b)$$

$$\int \cos(ax + b)\, dx = \frac{1}{a} \sin(ax + b)$$

$$\int \sin^2 x\, dx = \int \tfrac{1}{2}(1 - \cos 2x)\, dx = \tfrac{1}{2}(x - \tfrac{1}{2}\sin 2x)$$

$$\int \cos^2 x\, dx = \int \tfrac{1}{2}(1 + \cos 2x)\, dx = \tfrac{1}{2}(x + \tfrac{1}{2}\sin 2x)$$

$$\int \frac{dx}{x^2 - a^2} = \frac{1}{2a} \ln \left| \frac{x - a}{x + a} \right|$$

$$\int \frac{dx}{\sqrt{a^2 - b^2 x^2}} = \frac{1}{b} \sin^{-1}\left(\frac{bx}{a}\right)$$

$$\int \frac{dx}{a^2 + b^2 x^2} = \frac{1}{ab} \tan^{-1}\left(\frac{bx}{a}\right)$$

General Integration Including Simple Areas

EXERCISE 29a

Integrate each of the given expressions with respect to x.

1 $5x - 4$

2 $(3x + 2)^3$

3 $\dfrac{1}{(3x - 1)^2}$

4 $x + \dfrac{1}{x^2}$

5 $(x^2 - 3)^2$

6 $1 + 2\sqrt{x}$

7 $\dfrac{1}{\sqrt{4 - 3x}}$

8 $\dfrac{5 + x^3}{2x^2}$

9 $x(1 + x^2)^4$

10 Find the area enclosed between the curve $y = x^2 - 5x + 4$ and the x-axis.

11 Show that the area between the curve $y = x(x - 1)(x - 2)$ and the x-axis from $x = 0$ to $x = 1$ is numerically equal to the area between the same curve and the x-axis from $x = 1$ to $x = 2$.

12 Draw, on the same diagram, sketches of the graphs of the curves with equations $y = x(1 - x)$ and $y = x(x - 1)(x - 2)$.

Find the area of the crescent shape bounded by these curves.

13 Find the area enclosed by the curve $y = x^2 - 4$ and the line $y = 5$.

14 Find the area enclosed by the curve $y = x^2 - 2x - 3$ and the line $y = x + 7$.

65

15 A curve passes through the point $(2, 0)$ and its gradient at the point (x, y) is $3x^2 - 12x + 8$ for all values of x. Find the equation of the curve.

Find also the area in the first quadrant bounded by the x-axis and the curve between $x = 0$ and $x = 2$.

Show that the straight line $y = 3x$ divides this area in the ratio $3:13$.

Trigonometric Functions

EXERCISE 29b
Find:

1 $\displaystyle\int \sin 3x \, dx$ **2** $\displaystyle\int \cos 4x \, dx$ **3** $\displaystyle\int \tan 2x \, dx$

4 $\displaystyle\int \sec^2 2x \, dx$ **5** $\displaystyle\int \sin^2 2x \, dx$ **6** $\displaystyle\int \cos^2 \frac{x}{2} \, dx$

7 $\displaystyle\int 3 \sin \left(2x - \frac{\pi}{4}\right) dx$ **8** $\displaystyle\int 4 \cos \left(3x + \frac{\pi}{6}\right) dx$ **9** $\displaystyle\int \sec 3x \tan 3x \, dx$

10 $\displaystyle\int \sin^2 3x \, dx$ **11** $\displaystyle\int \sin^2 \frac{x}{3} \, dx$ **12** $\displaystyle\int \sin^2 x \cos x \, dx$

13 $\displaystyle\int \cos^3 x \sin x \, dx$ **14** $\displaystyle\int \sin^3 x \cos^3 x \, dx$ **15** $\displaystyle\int \frac{\cos x}{\sin^2 x} \, dx$

16 $\displaystyle\int_0^{\frac{\pi}{4}} \sin 2x \, dx$ **17** $\displaystyle\int_0^{\frac{\pi}{6}} \cos 3x \, dx$ **18** $\displaystyle\int_0^{\frac{\pi}{4}} \sin \left(\frac{\pi}{3} - \frac{x}{3}\right) dx$

19 $\displaystyle\int_{\frac{\pi}{6}}^{\frac{\pi}{3}} \sin x \cos^2 x \, dx$ **20** $\displaystyle\int_{\frac{\pi}{12}}^{\frac{\pi}{6}} \text{cosec}^2 2x \, dx$ **21** $\displaystyle\int_{-\frac{\pi}{2}}^{\frac{\pi}{2}} \cos \frac{x}{2} \, dx$

22 $\displaystyle\int_0^{\frac{\pi}{4}} \tan^2 x \, dx$ **23** $\displaystyle\int_0^{\frac{\pi}{2}} \sin^2 x \cos^2 x \, dx$ **24** $\displaystyle\int_0^{\frac{\pi}{6}} \cos^3 x \, dx$

25 Show that $\displaystyle\int_0^{\frac{\pi}{4}} \sin 2x \sin x \, dx = \frac{\sqrt{2}}{6}$

26 Show that $\displaystyle\int_{\frac{\pi}{3}}^{\frac{\pi}{2}} \sin 2x \sqrt{\cos x} \, dx = \frac{\sqrt{2}}{10}$

Logarithmic and Exponential Functions

EXERCISE 29c

Find:

1 $\displaystyle\int \frac{dx}{x+1}$

2 $\displaystyle\int \frac{3}{2x+1}\,dx$

3 $\displaystyle\int \frac{x}{1+x^2}\,dx$

4 $\displaystyle\int \frac{x^2}{1-x^3}\,dx$

5 $\displaystyle\int \frac{2x+3}{x^2+3x-4}\,dx$

6 $\displaystyle\int \frac{x-2}{x^2-4x+7}\,dx$

7 $\displaystyle\int e^{5x}\,dx$

8 $\displaystyle\int e^{-3x}\,dx$

9 $\displaystyle\int \frac{1}{e^{2x}}\,dx$

10 $\displaystyle\int \frac{e^x}{1+e^x}\,dx$

11 $\displaystyle\int \frac{2}{1+e^x}\,dx$

12 $\displaystyle\int \frac{1}{\sqrt{e^{3x}}}\,dx$

13 $\displaystyle\int_2^4 \frac{dx}{5x+1}$

14 $\displaystyle\int_0^1 \frac{1}{e^{5x}}\,dx$

15 $\displaystyle\int_0^{\frac{2}{3}} e^{2-3x}\,dx$

Areas and Volumes

EXERCISE 29d

1 Sketch the graphs of $y^2 = 4x$ and $y^2 = 8(x-2)$ showing clearly the points where the two graphs intersect.

Find the area enclosed between these two graphs and the volume generated when this area is turned through half a revolution about the x-axis.

2 Show that the area bounded by the y-axis, the curve $y = \sin x$ and the curve $y = \cos x$, for $0 \leqslant x \leqslant \frac{\pi}{4}$, is $\sqrt{2} - 1$ square units.

Find the volume generated when this area is rotated through a complete revolution about the x-axis.

3 Sketch the curve $y = \sin x$ for $0 \leqslant x \leqslant 2\pi$. Use this to sketch, on separate axes, for $0 \leqslant x \leqslant 2\pi$, the curves:

a) $y = \operatorname{cosec} x$ b) $y = \operatorname{cosec} x + \sin x$ c) $y = \operatorname{cosec} x - \sin x$

4 Sketch the curve $y = \cos x - \sin x$ for $0 \leqslant x \leqslant \frac{\pi}{2}$, showing clearly where it crosses the x-axis.

Find the area enclosed between the curve and the positive x and y axes, within the given range. If this area is rotated about the x-axis through four right angles, show that the volume generated is $\frac{\pi}{4}(\pi - 2)$ cubic units.

5 Sketch the curve $y^2 = x^2(4-x)$. The loop is rotated about the x-axis to form a solid of revolution. Find the volume of the solid formed.

67

6 Sketch, in separate diagrams, the graphs of $y = e^x$ and $y = \ln x$.

Find the area between the axes, the curve $y = e^x$ and the ordinate at $x = 1$. What is the volume generated when this area is rotated, through four right angles, about the x-axis?

Write down, without performing an integration, the volume generated when the area between the axes, the curve $y = \ln x$ and the line $y = 1$ is rotated about the y-axis. Deduce that the volume generated when the region bounded by the x-axis, the curve $y = \ln x$ and the ordinate at $x = e$ is rotated about the y-axis, is $\frac{\pi}{2}(1 + e^2)$.

Inverse Trigonometric Functions

EXERCISE 29e

Find:

1 $\displaystyle\int \frac{dx}{\sqrt{4 - x^2}}$
2 $\displaystyle\int \frac{dx}{\sqrt{9 - x^2}}$
3 $\displaystyle\int \frac{dx}{\sqrt{9 - 4x^2}}$

4 $\displaystyle\int \frac{dx}{x^2 + 9}$
5 $\displaystyle\int \frac{dx}{x^2 + 16}$
6 $\displaystyle\int \frac{dx}{4 + x^2}$

7 $\displaystyle\int \frac{dx}{x^2 + 2x + 5}$
8 $\displaystyle\int \frac{dx}{\sqrt{3 - 2x - x^2}}$
9 $\displaystyle\int \frac{dx}{\sqrt{1 - 4x^2}}$

10 $\displaystyle\int_{-2}^{2} \frac{dx}{x^2 + 4}$
11 $\displaystyle\int_{-1}^{1} \frac{dx}{\sqrt{2 - x^2}}$
12 $\displaystyle\int_{-3}^{-2} \frac{dx}{(x + 3)^2 + 1}$

13 $\displaystyle\int_{0}^{\frac{2}{3}} \frac{dx}{\sqrt{4 - 9x^2}}$
14 $\displaystyle\int_{\frac{2}{3}}^{1} \frac{dx}{9x^2 + 4}$
15 $\displaystyle\int_{1}^{\frac{3}{2}} \frac{3}{\sqrt{9 - 4x^2}}\,dx$

Integration by Substitution

EXERCISE 29f

Use the given substitution to find the integrals in questions 1 to 10.

1 $\displaystyle\int \frac{x}{(1 - 5x^2)^4}\,dx$, $\quad 1 - 5x^2 = u$

2 $\displaystyle\int x(1 + 3x^2)^5\,dx$, $\quad x^2 = u$

3 $\displaystyle\int x\sqrt{x^2 + 3}\,dx$, $\quad x^2 + 3 = u^2$

4 $\displaystyle\int \frac{x\,dx}{\sqrt{x^2 + 3}}$, $\qquad x^2 + 3 = u^2$

5 $\displaystyle\int x^3\sqrt{x^2 + 1}\,dx$, $\qquad x^2 + 1 = u^2$

6 $\displaystyle\int \frac{dx}{1 + \cos x}$, $\qquad \tan\frac{x}{2} = t$

7 $\displaystyle\int \frac{dx}{1 + \sin x}$, $\qquad \tan\frac{x}{2} = t$

8 $\displaystyle\int_{-2}^{1} \frac{dx}{x^2 + 4x + 13}$, $\qquad x + 2 = z$

9 $\displaystyle\int_{2}^{4} \frac{x(x^2 + 4)}{x^2 - 2}\,dx$, $\qquad x^2 - 2 = y$

10 $\displaystyle\int_{-1}^{3} \frac{4\,dx}{x^2 + 2x + 17}$, $\qquad x + 1 = z$

11 Use the substitution $t = \tan\dfrac{x}{2}$ to show that:

a) $\displaystyle\int_{0}^{\frac{\pi}{2}} \frac{dx}{5 + 3\cos x} = \tfrac{1}{2}\tan^{-1}\tfrac{1}{2}$

b) $\displaystyle\int_{0}^{\frac{\pi}{3}} \frac{dx}{2 - \cos x} = \frac{\pi}{2\sqrt{3}}$

12 Use the substitution $y = \ln x$ to evaluate $\displaystyle\int_{e}^{e^3} \frac{dx}{x \ln x}$.

Integration Involving Partial Fractions

EXERCISE 29g
Find:

1 $\displaystyle\int \frac{dx}{(x + 4)(x - 1)}$ $\qquad$ **2** $\displaystyle\int \frac{(x + 3)}{x(x + 1)}\,dx$ $\qquad$ **3** $\displaystyle\int \frac{25\,dx}{(x + 1)(x - 4)^2}$

4 $\displaystyle\int \frac{x^2 + 3}{x(x^2 + 2)}\,dx$ $\qquad$ **5** $\displaystyle\int \frac{x + 2}{x(x^2 + 1)}\,dx$ $\qquad$ **6** $\displaystyle\int \frac{4}{x^2 - 9}\,dx$

7 $\displaystyle\int \frac{x^2}{(x + 1)(x - 1)}\,dx$ $\qquad$ **8** $\displaystyle\int \frac{dx}{(x + 2)^2 - 9}$ $\qquad$ **9** $\displaystyle\int \frac{x^2 - 1}{(x - 3)^3}\,dx$

10 $\displaystyle\int_{2}^{3} \frac{dx}{x^2 + x}$ $\qquad$ **11** $\displaystyle\int_{0}^{2} \frac{x}{(1 + x)^2}\,dx$ $\qquad$ **12** $\displaystyle\int_{2}^{5} \frac{x\,dx}{(x - 4)(x - 1)}$

69

13 Express $\dfrac{3}{x^2(x+3)}$ in partial fractions, and hence show that

$$\int_2^4 \dfrac{3}{x^2(x+3)}\,dx = \tfrac{1}{12}(3 + 4\ln\tfrac{7}{10}).$$

14 Express $\dfrac{1}{x^2(x+1)}$ in partial fractions, and hence show that

$$\int_1^2 \dfrac{dx}{x^2(x+1)} = 0.212 \text{ (correct to 3 s.f.).}$$

Integration by Parts

EXERCISE 29h

Use integration by parts to find:

1 $\displaystyle\int x\,e^x\,dx$ **2** $\displaystyle\int \ln x\,dx$ **3** $\displaystyle\int x\ln x\,dx$

4 $\displaystyle\int x\sin x\,dx$ **5** $\displaystyle\int x^2\cos x\,dx$ **6** $\displaystyle\int xe^{5x}\,dx$

7 $\displaystyle\int xe^{-2x}\,dx$ **8** $\displaystyle\int \sin^{-1} x\,dx$ **9** $\displaystyle\int x\tan^{-1} x\,dx$

10 $\displaystyle\int x^2 e^x\,dx$ **11** $\displaystyle\int x^2 \ln x\,dx$ **12** $\displaystyle\int e^{2x}\cos x\,dx$

13 If $I_n = \displaystyle\int x^n\,e^x\,dx$ use integration by parts to show that $I_n = x^n\,e^x - nI_{n-1}$ $(n \geqslant 1)$.

Use this reduction formula to find $\displaystyle\int x^3\,e^x\,dx$ and hence find the value of $\displaystyle\int_0^1 x^3\,e^x\,dx$.

14 Show that $\displaystyle\int \tan^2 x\,dx = \tan x - x$. If $I_n = \displaystyle\int \tan^n x\,dx$ $(n \geqslant 2)$ show that

$$I_n + I_{n-2} = \int \tan^{n-2} x\sec^2 x\,dx = \dfrac{1}{n-1}\tan^{n-1} x.$$ Hence find $\displaystyle\int \tan^6 x\,dx$.

The Trapezium Rule

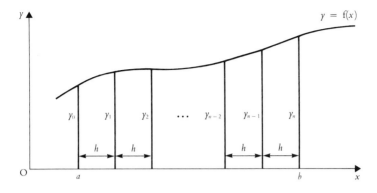

The area between the curve $y = f(x)$, the x-axis and the ordinates at $x = a$ and $x = b$ is given by:

$$\int_a^b f(x)\,dx \approx \frac{h}{2}[y_0 + 2y_1 + 2y_2 + \ldots + 2y_{n-1} + y_n]$$

where n is the number of strips.

EXERCISE 29i

In questions 1 to 6 find the approximate values of the integrals using the trapezium rule and the number of strips indicated in brackets. Remember that if there are n strips there will be $(n + 1)$ ordinates.

1 $\int_1^7 \frac{1}{x}\,dx$ (6)

2 $\int_0^1 e^x\,dx$ (5)

3 $\int_0^{\frac{3}{2}} e^{x^2}\,dx$ (6)

4 $\int_0^{\frac{\pi}{2}} \sin x\,dx$ (6)

5 $\int_0^{0.8} \sqrt{1 + x}\,dx$ (8)

6 $\int_0^{0.7} \ln(1 + x)$ (7)

Simpson's Rule

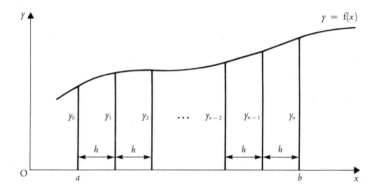

The area between the curve $y = f(x)$, the x-axis and the ordinates at $x = a$ and $x = b$ is given by:

$$\int_a^b f(x)\,dx \approx \frac{h}{3}\,[y_0 + 4(\text{odd ordinates}) + 2(\text{even ordinates}) + y_n]$$

or $\dfrac{h}{3}\,[(y_0 + y_n) + 4(y_1 + y_3 + \ldots) + 2(y_2 + y_4 + \ldots)]$

Note that n must always be even i.e. there must be an even number of strips.

EXERCISE 29j

In questions 1 to 6 find the approximate values of the integrals using Simpson's Rule and the number of strips indicated in brackets.

1 $\displaystyle\int_0^{\frac{\pi}{3}} \sin x\,dx$ (6)

2 $\displaystyle\int_0^{\frac{\pi}{2}} \sqrt{\cos x}\,dx$ (6)

3 $\displaystyle\int_0^1 \frac{1}{1 + x^2}\,dx$ (8)

4 $\displaystyle\int_0^{1.2} \ln (1 + x)\,dx$ (6)

5 $\displaystyle\int_0^{0.6} x\,e^x\,dx$ (6)

6 $\displaystyle\int_{\frac{\pi}{4}}^{\frac{\pi}{2}} \frac{1}{1 + \sin x}\,dx$ (4)

Numerical Solutions to Equations

30

Iterative Methods

For an iterative method to give a satisfactory solution the iterates, or successive approximate solutions, must converge i.e. a stage must be reached when successive iterates are equal. Much depends on a good first approximation.

EXERCISE 30a

1 Use the iterative formula $x_{r+1} = 8 - \dfrac{3}{x_r}$ to find the larger root of the quadratic equation $x^2 - 8x + 3 = 0$. Start with $x_1 = 8$, and find x_2, x_3 and x_4.

2 The three roots of the equation $x^3 - 8x + 3 = 0$ are approximately -3, 0.3 and 2.6. The iterative formula $x_{r+1} = \dfrac{x_r^3 + 3}{8}$ can be used to find just one of the roots more accurately. Find this root correct to four decimal places by working as far as x_4.

3 Two possible iterative formulae to solve the equation $x^3 - 20 = 0$ are
$$x_{r+1} = \sqrt{\frac{20}{x_r}} \text{ and } x_{r+1} = \frac{2x_r}{3} + \frac{20}{3x_r^2}.$$

Starting from $x_1 = 3$, find x_2, x_3 and x_4 for each of these formulae. Comment on your results. Hence find $\sqrt[3]{20}$ correct to four decimal places.

4 Use the iterative formula $x_{r+1} = 1 - \frac{1}{2}\cos x_r$ to solve the equation $2x + \cos x = 2$ correct to four decimal places. Start with $x_1 = 0.6$

Newton–Raphson

If h is an approximate root of the equation $f(x) = 0$ then, in general, $h - \dfrac{f(h)}{f'(h)}$ is a better approximation. This method will lead to a root of the equation $f(x) = 0$ provided $f'(x) \neq 0$ near the exact root.

EXERCISE 30b

1 Show that $3x^3 - 2x^2 - 7x + 4 = 0$ has a root between 1 and 2, and find this root correct to two decimal places.

2 Show that $x^3 - 6x + 7 = 0$ has a root between -3 and -2, and find this root correct to two decimal places.

Use a graphical method to find a first approximation to the positive root of the given equations in questions 3 to 7. Apply the Newton–Raphson method twice to give a better approximation.

3 $\ln(1 + x) = \dfrac{1}{x}$

4 $e^x = 2 - 3x$

5 $\sin x = 1 - x$

6 $\cos x = 2x$

7 $1 + \sin x = x$

8 Show that the equation $e^x = 1 + \sin x + \cos x$ has a root between -4 and -3. Find this root correct to three decimal places.

Complex Numbers $\quad$ **31**

If $z = a + bi$ the conjugate of z, which is denoted by $\bar{z}$, is $a - bi$.

The complex roots of equations with real coefficients occur in conjugate pairs.

If two complex numbers are equal their real and imaginary parts are separately equal, i.e. if $a + bi = c + di$ then $a = c$ and $b = d$.

The modulus of the complex number $z = a + bi$ is denoted by $|z|$ where $|z| = \sqrt{a^2 + b^2}$.

The principal argument of z, denoted by $\arg z$, is the angle θ $(-\pi < \theta < \pi)$ such that $\tan \theta = \dfrac{b}{a}$ i.e. $\arg z = \tan^{-1}\dfrac{b}{a}$ or $\arctan\dfrac{b}{a}$.

The argument gives the direction, from the origin to the point in the Argand diagram representing z, measured from the positive x-axis.

In the Argand diagram, if $z = x + yi$ is represented by $\overrightarrow{OP}$, where P has Cartesian coordinates (x, y) and the corresponding polar coordinates of P are (r, θ), then:

$$x = r\cos\theta, \quad y = r\sin\theta, \quad r^2 = x^2 + y^2$$

and $\qquad\qquad \tan\theta = \dfrac{y}{x}$

Hence $\qquad\qquad x + yi = r\cos\theta + ir\sin\theta$

$$= r\operatorname{cis}\theta$$

If two complex numbers are multiplied together we multiply their moduli but add their arguments. If one complex number is divided by another complex number we divide their moduli but substract their arguments. Hence if $z_1 = r_1 \operatorname{cis} \theta_1$ and $z_2 = r_2 \operatorname{cis} \theta_2$:

$$z_1 z_2 = r_1 r_2 \operatorname{cis} (\theta_1 + \theta_2)$$

and

$$\frac{z_1}{z_2} = \frac{r_1}{r_2} \operatorname{cis} (\theta_1 - \theta_2)$$

Some well-known geometrical relations can be expressed as equations involving complex numbers.

a)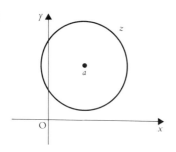

$|z - a| = r$
A circle, centre at the point a, radius r.

b)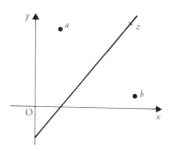

$|z - a| = |z - b|$
The perpendicular bisector of the straight line joining the points represented by the complex numbers a and b.

c)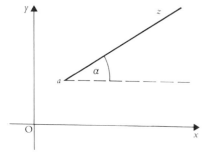

$\arg (z - a) = \alpha$
The line from the point representing a in the direction α.

EXERCISE 31

1 Find the sum, difference, product and quotient of each of the following pairs of complex numbers. Give each answer in the form $a + bi$.

a) $3 + i, 2 - i$

b) $-1 + i, 2 - 3i$

c) $3i, 4i$

d) $i - 3, i + 4$

2 Simplify, giving each answer in the form $a + bi$.

a) $(i + 2)^2 + (i - 3)^2$

b) $(1 + 2i)^3$

c) $(1 + i)^3 - (1 - i)^3$

d) $3 + 2i - \dfrac{1}{2 + i}$

e) $\dfrac{3}{3 - 2i}$

f) $\dfrac{1}{2 + i} - \dfrac{1}{2 - i}$

g) $\dfrac{(2 - i)^2}{3 + i}$

h) $\dfrac{4 + 3i}{2 - i}$

3 Solve the equations:

a) $x^2 + 4x + 5 = 0$

b) $x^2 - 2x + 4 = 0$

c) $x^2 + 6x + 10 = 0$

d) $2x^2 + 5x + 4 = 0$

e) $4x^2 - 12x + 13 = 0$

f) $4x^2 - 8x + 5 = 0$

4 Find all the roots for each of the following equations.

a) $x^3 + x = 0$

b) $x^3 - 5x^2 + 9x - 5 = 0$

c) $x^3 + x^2 - 2 = 0$

d) $x^3 - 8x^2 + 25x - 26 = 0$

5 Verify that $z = i$ satisfies the equation $z^4 + 4z^3 + 6z^2 + 4z + 5 = 0$ and find the other roots.

6 Verify that $z = -3 + 2i$ satisfies the equation $z^4 + 6z^3 + 17z^2 + 24z + 52 = 0$ and find the other roots.

7 If $\dfrac{a + bi}{c + di} = x + yi$ find x and y in terms of a, b, c and d.

8 If $z = 1 + i$ mark in the Argand diagram the points that represent z, iz, $i^2 z$ and $i^3 z$. Show that these points are the vertices of a square and find the length of a side of this square. Write down the modulus and argument of each of the four complex numbers represented on this diagram.

9 Find the modulus and argument of each of these complex numbers (assume that α is acute).

a) $1 + 2i$

b) $1 - \sqrt{3}i$

c) $2 + \sqrt{3}i$

d) $1 + \dfrac{i}{\sqrt{3}}$

e) $-\sqrt{3} + 2i$

f) $-2 - i$

g) $\cos \alpha + i \sin \alpha$

h) $1 + i \tan \alpha$

i) $1 + \cos \alpha - i \sin \alpha$

10 If the point A in the Argand diagram represents the complex number $4 - 5i$, write down the complex number that represents:

a) a reflection of A in the x-axis,

b) a reflection of A in the y-axis,

c) a reflection of A in the line $y = x$,

d) a reflection of A in the line $y = -x$.

11 Express in the form $x + yi$ the complex number represented on the Argand diagram by $\overrightarrow{OP}$ where the polar coordinates of P are:

a) $(2, \frac{\pi}{4})$ b) $(4, \frac{\pi}{6})$ c) $(1, -\frac{\pi}{3})$

d) $(4, -\frac{2\pi}{3})$ e) $(2, -\frac{\pi}{2})$ f) $(2, \frac{3\pi}{4})$

12 Find the Cartesian equation of the locus of the points in the Argand diagram such that:

a) $|z| = 2$ b) $|z - 1| = 3$

c) $|z + i| = 3$ d) $|z - 1| = |z + 1|$

e) $|z - 2| = |z - 4|$ f) $|z + i| = |z - 2|$

13 If $z = 2 - 3i$ evaluate $\bar{z}$, $z + 4$ and $\bar{z} - 4$. Plot points, to represent these four complex numbers, in the Argand diagram. Interpret these results geometrically.

14 If $(a + bi)^2 = x + yi$ show that $a^2 - b^2 = x$ and $2ab = y$. Hence evaluate $\sqrt{5 + 12i}$.

15 If $z = a + bi$ express: a) $\frac{1}{z}$,   b) $\bar{z}$,   c) $\frac{1}{\bar{z}}$, in the form $x + yi$.

16 If $z = r(\cos\theta + i\sin\theta)$ express: a) $\frac{1}{z}$,   b) $\bar{z}$,   c) $\frac{1}{\bar{z}}$, in the form $x + yi$.

17 ABCD is a square described in an anticlockwise sense. If A and B respectively represent $4 - 2i$ and $3 + 2i$, find the complex numbers represented by C and D.

18 The complex numbers u, v and w are such that $\frac{1}{u} + \frac{1}{v} = \frac{1}{w}$. If $u = 2 + 3i$ and $v = 3 + 2i$ find w in the form $a + bi$.

19 PQRS is a rectangle described in an anticlockwise sense. $QR = 2PQ$. If P and Q represent $-2 - 3i$ and $-3 + 2i$ respectively, find the numbers represented by R and S.

20 Show on an Argand diagram where the point z must lie if it satisfies the following condition.

a) $|z - 2| = 2$ b) $|z - 1| > 2$

c) $|z + 2 - i| = 1$ d) $|z - i| < 1$

e) $\arg z = \frac{\pi}{2}$ f) $\arg(z - 1) = \frac{\pi}{6}$

g) $\arg z = \pi$ h) $\arg(z - i) = \frac{\pi}{3}$

21 ABCD is a square and the points B and D represent the numbers $-2 + 2i$ and $4 - 2i$. What number represents the midpoint of BD? Find the numbers represented by A and C.

22 Find the square root of: a) $8 - 6i$ b) $21 + 20i$ c) $7 - 24i$ d) $30i - 16$

23 Form the equation whose roots are:

a) $3 + i$, $3 - i$

b) $2 + 3i$, $2 - 3i$

c) 1, $1 + i$, $1 - i$

d) -2, $4 + 3i$, $4 - 3i$

24 Express in the form $r(\cos \theta + i \sin \theta)$:

a) 3

b) $4i$

c) $1 + i$

d) $\sqrt{3} + i$

e) $3 - i\sqrt{3}$

f) $-3 - 4i$

25 Draw separate diagrams to show the values of z that satisfy each of the following inequalities in the Argand plane.

a) $R(z) \geqslant 0$

b) $0 \leqslant \arg z \leqslant \frac{\pi}{3}$

c) $|z| \leqslant |z - 1|$

Combine these conditions into a single diagram to show the region in the Argand plane that satisfies all the above inequalities.

26 Solve the equations:

a) $z^2 - (3 + 2i)z + 1 + 3i = 0$

b) $z^2 + (1 - 3i)z - (8 - i) = 0$

c) $z^2 - (4 + 5i)z - 3 + 9i = 0$

Vectors

32

The modulus of the vector:

$$\mathbf{r} = a\mathbf{i} + b\mathbf{j} + c\mathbf{k}$$

is given by:

$$|\mathbf{r}| = \sqrt{a^2 + b^2 + c^2}$$

The direction ratios of this vector are $a:b:c$ while the direction cosines of this vector are:

$$\frac{a}{\sqrt{a^2 + b^2 + c^2}}, \quad \frac{b}{\sqrt{a^2 + b^2 + c^2}}, \quad \frac{c}{\sqrt{a^2 + b^2 + c^2}}$$

The sum of the squares of the direction cosines of any vector is unity.

The equation of the straight line through the point (x_1, y_1, z_1) that is parallel to the vector $a\mathbf{j} + b\mathbf{j} + c\mathbf{k}$ is given by:

$$\mathbf{r} = x_1\mathbf{i} + y_1\mathbf{j} + z_1\mathbf{k} + \lambda(a\mathbf{i} + b\mathbf{j} + c\mathbf{k})$$

or

$$\frac{x - x_1}{a} = \frac{y - y_1}{b} = \frac{z - z_1}{c} = (\lambda)$$

or

$$x = x_1 + \lambda a, \quad y = y_1 + \lambda b, \quad z = z_1 + \lambda c$$

where $a:b:c$ are the direction ratios of the given vector. The equation $\mathbf{r} = \mathbf{a} + \lambda(\mathbf{b} - \mathbf{a})$ is the equation of the straight line passing through the two points whose position vectors are $\mathbf{a}$ and $\mathbf{b}$.

The angle between the two lines whose equations are:

$$r_1 = x_1i + y_1j + z_1k + \lambda(a_1i + b_1j + c_1k)$$
$$r_2 = x_2i + y_2j + z_2k + \mu(a_2i + b_2j + c_2k)$$

is given by:

$$\cos\theta = \frac{a_1a_2 + b_1b_2 + c_1c_2}{|r_1||r_2|}$$

where $a_1:b_1:c_1$ and $a_2:b_2:c_2$ are the respective direction ratios of the given lines.

EXERCISE 32

1 Find the coordinates of P, where $|\overrightarrow{OP}| = 14$ and $\overrightarrow{OP}$ is in the direction $2i + 6j - 3k$.

2 A, B and C are three points whose position vectors are respectively $3i + j - 2k$, $i + 2j + 3k$, $2i - j + k$. P and Q are the midpoints of AB and AC. Find the vectors $\overrightarrow{BC}$ and $\overrightarrow{PQ}$. Hence show that PQ is parallel to BC and equal to one-half of it.

3 Write down the direction cosines of the line with equation:

$$r = \begin{pmatrix} 1 \\ 2 \\ 3 \end{pmatrix} + \lambda \begin{pmatrix} 2 \\ -2 \\ 1 \end{pmatrix}$$

4 Convert the vector equation $r = i + 2j - 3k + \lambda(2i + j + 3k)$ into Cartesian form.

5 State whether or not the lines whose equations are given below are parallel.

$$\frac{x-1}{3} = \frac{2-y}{1} = \frac{z-3}{2}$$

$$\frac{x-2}{6} = -\frac{y}{2} = \frac{z-3}{4}$$

6 State whether or not the lines:

$$r = i + 2j - 3k + \lambda(2i + j - k)$$

and

$$r = 2i + 4j - 6k + \lambda(-4i - 2j + k)$$

are parallel.

7 A line has Cartesian equations $\frac{x-1}{4} = \frac{y+2}{3} = \frac{z-3}{2}$. Find a vector equation for a parallel line passing through the point with position vector $3i + 4j - 2k$ and find the coordinates of the point where this line crosses the xy plane.

79

8 Show that the lines whose vector equations are:

$$\mathbf{r}_1 = \mathbf{i} + \mathbf{j} + 3\mathbf{k} + \lambda(3\mathbf{i} + \mathbf{j} + 2\mathbf{k})$$

and $$\mathbf{r}_2 = 2\mathbf{i} + 4\mathbf{j} + \mathbf{k} + \mu(-\mathbf{i} + \mathbf{j} - 2\mathbf{k})$$

intersect, and find the position vector of their point of intersection.

9 Show that the two straight lines given by the vector equations:

$$\mathbf{r}_1 = 3\mathbf{i} + 2\mathbf{j} + \mathbf{k} + \lambda(2\mathbf{i} - 3\mathbf{j} + \mathbf{k})$$

and $$\mathbf{r}_2 = \mathbf{i} - 3\mathbf{j} + 2\mathbf{k} + \mu(\mathbf{i} - 2\mathbf{j} + 3\mathbf{k})$$

are skew lines, i.e. they are neither parallel nor do they intersect.

10 Find the angle between the straight lines whose equations are:

$$\frac{x+3}{2} = \frac{y-1}{4} = \frac{z+4}{2}$$

and $$\frac{x+1}{1} = \frac{y-3}{1} = \frac{z-2}{2}$$

11 Show that $2\mathbf{i} - 3\mathbf{j} + 2\mathbf{k}$ is perpendicular to both $\mathbf{i} + 2\mathbf{j} + 2\mathbf{k}$ and $3\mathbf{i} - 4\mathbf{j} - 9\mathbf{k}$.

12 The Cartesian equations of a line are:

$$\frac{1-x}{4} = y = \frac{z+2}{3}$$

Convert these into the vector equation of the line, and hence find the direction cosines of this line.

13 Find the acute angle between the two straight lines whose vector equations are:

$$\mathbf{p} = \mathbf{i} + 2\mathbf{k} + \lambda(5\mathbf{i} + 3\mathbf{j} - 2\mathbf{k})$$

and $$\mathbf{q} = 2\mathbf{i} - \mathbf{j} + 3\mathbf{k} + \mu(-2\mathbf{i} + 3\mathbf{j} + 5\mathbf{k})$$

14 The angle between the two straight lines whose vector equations are:

$$\mathbf{p} = 2\mathbf{i} + \mathbf{j} - 3\mathbf{k} + \lambda(a\mathbf{i} - 2\mathbf{j} + 4\mathbf{k})$$

and $$\mathbf{q} = \mathbf{i} + \mathbf{j} + 2\mathbf{k} + \mu(\mathbf{i} + 2\mathbf{j} + 2\mathbf{k})$$

is $\cos^{-1}\frac{4}{9}$. Find the integral value of a.

Differential Equations where the Variables Separate

33

EXERCISE 33

Find the general solution of the following differential equations.

1 $\dfrac{dy}{dx} = \dfrac{x}{y}$

2 $\dfrac{dy}{dx} = \dfrac{y}{x}$

3 $\dfrac{dy}{dx} = 4ax^3$

4 $3x\dfrac{dy}{dx} = 4y$

5 $(x + 4)\dfrac{dy}{dx} = y$

6 $\dfrac{dy}{dx} = xy$

7 $\dfrac{dy}{dx} = \dfrac{y + 2}{x - 3}$

8 $y\dfrac{dy}{dx} = \dfrac{x}{\sqrt{x^2 - 1}}$

9 $x(x + 1)\dfrac{dy}{dx} = y^2$

10 $(x^2 + 1)\dfrac{dy}{dx} = 2xy$

11 $\dfrac{dy}{dx} = e^{2x}(y^2 - 1)$

12 $\dfrac{dv}{du} = v(v + 1)$

13 $x \ln x\dfrac{dy}{dx} = y$

14 $2\dfrac{dy}{dx} = e^x \sin 2y$

15 $\dfrac{x^2}{y}\dfrac{dy}{dx} = \ln x$

Part 2: Revision Papers 1-10

Revision Paper 1

1 If $\tan \theta = \frac{3}{4}$ and θ is acute, find the value of:

   a) $\sin 2\theta$,   b) $\cos 2\theta$,   c) $\tan 2\theta$

2 Prove that $\cot \theta + \tan \theta = \operatorname{cosec} \theta \sec \theta$.

3 Find the smallest angle in a triangle whose sides are of lengths 5.4 cm, 6.8 cm and 7.6 cm. What is the area of the triangle?

4 Find the smallest value of n so that:

$$1 + 2 + 3 + 4 + \ldots + n > 500$$

5 The equation of a curve is $x^2 + 4y^2 = 8$. Show that the gradient of the tangent at the point $(2, 1)$ is $-\frac{1}{2}$. Hence find the equation of the normal at this point.

6 A sphere, originally 10 cm^3 in volume, is growing at the rate of 5 cm^3/min. Find the rate of increase of the radius after 5 minutes.

7 Solve the simultaneous equations $x^2 + y^2 + 3x + 7y - 40 = 0$, $y = x + 1$.

8 If the coefficient of x^3 in the binomial expansion of $(1 + 2x)^n$ is 448 find the positive integer value of n.

9 Show that the circles with equations:

$$x^2 + y^2 + 4x - 4y + 3 = 0$$

and $\qquad x^2 + y^2 - 10x - 2y - 19 = 0$

intersect at two points and find the coordinates of each.

Find the equation of the tangent to each circle at the point of intersection with the larger y-coordinate and show that these tangents are mutually perpendicular.

10 Show that the point $P\left(ap, \dfrac{a}{p}\right)$ lies on the rectangular hyperbola $xy = a^2$ for all values of p. Find the equation of the tangent at P.

If Q is the point $\left(aq, \dfrac{a}{q}\right)$ show that the tangents at P and Q intersect at the

point $\left(\dfrac{2apq}{p + q}, \dfrac{2a}{p + q}\right)$.

11 Express $x^2 + 2x + 5$ in the form $(x + a)^2 + b^2$. Hence find the minimum value of $x^2 + 2x + 5$ and the value of x for which this minimum value occurs.

Show that:

$$\int_0^1 \frac{dx}{x^2 + 2x + 5} = \frac{\pi}{8} - \frac{1}{2} \tan^{-1} \frac{1}{2}$$

12 Show that the volume of the largest cone that will fit inside a hollow sphere of radius R is $\frac{32}{81} \pi R^3$.

Revision Paper 2

1 By means of the substitution $y = 3^x$, or otherwise, find the two values of x such that:

$$3 . 3^x + 3^{-x} = 4$$

2 Three numbers are in GP. Their sum is 14 and their product is 64. Find them.

3 Given that $\sin(\theta - 30°) = 2 \cos(\theta + 60°)$ find the value of $\tan \theta$.

4 Factorise $\alpha^3 + \beta^3$.

If α and β are the roots of the quadratic equation $2x^2 - 6x + 1 = 0$ find the value of:

a) $\alpha^2 + \beta^2$ b) $\dfrac{1}{\alpha} + \dfrac{1}{\beta}$ c) $\dfrac{1}{\alpha^2} + \dfrac{1}{\beta^2}$ d) $\alpha^3 + \beta^3$

5 Show that the equation $x^2 + y^2 + 4x + 6y - 51 = 0$ represents a circle. Find its radius and the coordinates of the centre.

Show that the point $(6, 3)$ lies outside this circle.

Show that the lines $x = 6$ and $y = 5$ are tangents to this circle.

Find the length of a tangent from the point $(6, 3)$ to the circle.

6 Given that $x - 3$ and $x + 3$ are factors of $x^3 + ax^2 + bx + 9$, find the values of a and b. Use these values to factorise the resulting expression completely.

7

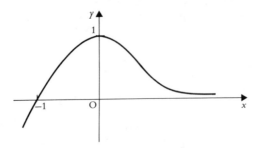

The sketch shows the graph of $y = f(x)$. The curve passes through the point $(-1, 0)$ and has a maximum point at $(0, 1)$. Sketch, on separate diagrams, the graphs of:

a) $y = f(x) + 1$ b) $y = f(x + 1)$ c) $y = f(2x)$

8 Use the trapezium rule with six strips to find the value of $\int_0^{0.6} xe^x \, dx$.

9 Find the coordinates of the point where the line through P $(4, 3, -2)$ and Q $(2, -1, 2)$ crosses the xy plane.

10 Find: a) $\dfrac{d}{dx}\left(\dfrac{e^{3x}}{\sin 2x}\right)$ b) $\displaystyle\int \dfrac{dx}{x(x + 1)}$

11 The curve whose equation is $y = x^3 - ax^2 + bx + c$ passes through the point $(1, 10)$, has a local minimum when $x = 3$ and a point of inflexion when $x = 2$. Find the values of a, b and c. Find the maximum value and sketch the curve.

12 A particle moves in a straight line so that its distance, x m, from a fixed point on the line, at time t seconds, is given by $x = t^3 - 6t^2 + 9t$.

a) When does the particle return to its starting point?
b) When is it instantaneously at rest?
c) What is its acceleration each time it is instantaneously at rest?
d) When is its acceleration zero, and what is its speed at this moment?

Revision Paper 3

1 Solve the simultaneous equations:

$$x^2 + y^2 = 25, \quad xy = 12$$

2 If $\sin \theta = \frac{4}{5}$ and θ is obtuse find the value of:

a) $\sin 2\theta$ b) $\cos 2\theta$ c) $\tan 2\theta$

3 Expand $\left(1 + \dfrac{x}{3}\right)^{-3}$ as far as the term in x^3. For what values of x is this expansion valid?

4 Prove that $\tan^{-1} 1 + \tan^{-1} \frac{1}{2} + \tan^{-1} \frac{1}{3} = \frac{\pi}{2}$.

5 Find the largest value of n so that

$$1 + 2 + 3 + 4 + \ldots + n < 1000.$$

6 The sum of the first 8 terms of a GP is 765, and the first term is 3. Find the common ratio by trial and error. Use your value of r to find the sum of the first 5 terms.

7 Show that $x^2 + 6x + 1 = (x + 3)^2 - 8$. Hence show that the minimum value of $x^2 + 6x + 1$ is -8. For what value of x does this minimum value occur?

8 Evaluate x and y where $z = x + yi$ if:

$$z(1 + 2i) = 2(3 - i)^2 - \frac{1}{4 + 3i}$$

9 PQ is a focal chord of the parabola $y^2 = 4ax$. If P is the point $(4a, 4a)$ show that the tangents at P and Q intersect at the point $T\left(-a, \dfrac{3a}{2}\right)$.

The normal to the parabola at P cuts the curve again at R. Show that R is the point $(9a, -6a)$.

If M is the midpoint of PQ and N is the midpoint of PR prove that both TM and QR are parallel to the axis of the parabola.

10 The perimeter of a sheet of metal, in the form of a sector of a circle, is 50 cm. What is the angle of the sector for which the area of the sector is greatest?

11 AB is a chord of a circle centre O which subtends an angle 2θ at O $(\theta < \frac{\pi}{2})$. If AB divides the minor sector of the circle into two equal parts prove that θ satisfies the equation $\theta = \sin 2\theta$. Show that θ lies between 0.94 and 0.95.

12 a) Show that $\displaystyle\int_0^{\frac{\pi}{6}} \sin 2x \cos^2 x \, dx = \frac{7}{32}$

b) Show that $\displaystyle\int \frac{dx}{1 + x + x^2} = \frac{2\sqrt{3}}{3} \tan^{-1} \frac{2x + 1}{\sqrt{3}}$

Revision Paper 4

1 If $x + y = a$ and $x^2 + y^2 = b^2$ find an expression for $(x^2 - y^2)^2$ in terms of a and b.

2 If $\cos \theta = \frac{2}{3}$ and θ is in the fourth quadrant find the value of:
a) $\sin 2\theta$,   b) $\cot 2\theta$.

3 Prove that $(\sin x + \cos x)^2 + (\sin x - \cos x)^2 = 2$.

4 Find the largest angle in, and the area of, the triangle whose sides are 9.7 cm, 7.1 cm and 6.3 cm.

5 Find the range of values of x for which $2x^2 + 3x > 9$.

6 Expand $(1 + x)^2 \sqrt{1 + 2x}$ in ascending powers of x as far as the term in x^3. For what values of x is the expansion valid?

7 Show that $x - 2\sqrt{x^2 + 1}$ has a maximum value at $x = \dfrac{1}{\sqrt{3}}$.

8 The normal at P to the parabola whose equation is $y^2 = 4ax$ meets the x-axis at N. If S is the point $(a, 0)$ prove that PSN is an isosceles triangle.

9 The points A and C represent in the Argand diagram the roots of the equation $z^2 - 6z + 13 = 0$. If AC is the diagonal of a square ABCD find:
a) the numbers represented by the points B and D, if D has the larger x-coordinate,
b) the area of the square ABCD.

10 What is the greatest volume of a right circular cone if the sum of its height and base radius is 20 cm?

11 Find the gradient of the tangent to the curve $y = \ln x$ at the point whose x-coordinate is a. Hence find the equation of the tangent drawn from the origin to the curve $y = \ln x$.

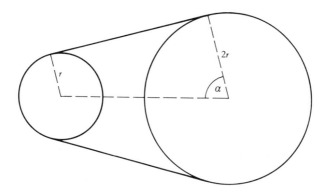

A belt passes around two wheels in the same plane as shown in the diagram. The radius of the smaller wheel is r and the radius of the larger wheel is $2r$. The centres are $4r$ apart.

Given that the belt is taut find:

a) the size of the angle marked α, in radians, correct to three significant figures,

b) the total length of the belt correct to the nearest whole number when $r = 5$ cm.

Revision Paper 5

1 Solve the equation $6\cos^2\theta - 5\cos\theta + 1 = 0$ for $0 \leqslant \theta \leqslant 360°$.

2 Prove by induction, or otherwise, that:

$$3 \times 4 + 5 \times 5 + 7 \times 6 + \ldots (2n + 1)(n + 3) = \frac{n}{6}(4n^2 + 27n + 41)$$

3 Prove that $\operatorname{cosec}^2\theta - \sec^2\theta = \cot^2\theta - \tan^2\theta$.

4 AB is a chord of a circle centre O, that subtends an angle θ $(<\pi)$ at O. If AB divides the circle into two parts, one having twice the area of the other, show that $3\theta - 3\sin\theta = 2\pi$. Show that $2.604 \leqslant \theta \leqslant 2.606$.

5 Find the values of a and b if $2x^2 + 8x + 3$ is expressed in the form $2(x + a)^2 + b$. Hence find the minimum value of $2x^2 + 8x + 3$ and the value of x for which it occurs.

6 Solve the simultaneous equations:

$$y = 2(x - 1), \quad 4x^2 - 2xy + y^2 = 12$$

7 Find the first four terms in the expansion of $(1 - x + 2x^2)^8$.

8 If $z = x + yi$ mark on the Argand diagram the points that represent the complex numbers z, $\bar{z}$, iz, $i\bar{z}$, $\dfrac{\bar{z}}{i}$ and $\dfrac{z}{i}$. Deduce the Cartesian equation of the curve on which all of these points lie.

9 The first three terms of a geometric progression are p, q, p^3 $(p > 0)$ and the first three terms of an arithmetic progression are p, p^2, $\frac{3}{2}q$.
a) Find the values of p and q.
b) Find expressions for the sum of n terms of each series.
c) By trial and error find a value for n such that the sum of n terms of the GP is exactly three times the sum of n terms of the AP.

10 The equations of two circles are:
$$x^2 + y^2 - 4x + 8y - 80 = 0$$
$$x^2 + y^2 + 14x - 16y + 88 = 0$$

Show that the point $(-4, 4)$ lies on each circle and find the equation of the tangent to each circle at this point.

Hence prove that the circles touch.

11 Express $\dfrac{x(x + 2)}{(x - 2)(x^2 + 4)}$ in partial fractions.

Hence find $\displaystyle\int \dfrac{x(x + 2)}{(x - 2)(x^2 + 4)}\, dx$.

12 The vertices of a triangle are P $(-2, -4)$, Q $(8, -2)$ and R $(-4, 6)$.
a) Find: i)   the gradient of PQ,
        ii) the coordinates of the midpoint, M, of PQ,
        iii) the equation of the perpendicular bisector of PQ.
b) Find the equation of the perpendicular bisector of PR.
c) Find the coordinates of T, the point in which the perpendicular bisectors of PQ and PR intersect.
d) Find the distance of T from each of the points P, Q and R.
e) Write down the coordinates of the centre of the circumcircle of triangle PQR.
f) The equation of the circle that passes through P, Q and R is $(x - a)^2 + (y - b)^2 = r^2$. Write down the values of a, b and r.

Revision Paper 6

1 Solve the simultaneous equations:
$$x - y = 3, \quad x^3 - y^3 = 117$$

2 Given that $3^x = u$, write down, in terms of u, the value of:
a) 3^{x+2},   b) 3^{2x}.

Hence solve the equation $3^{2x} - 3^{x+2} + 18 = 0$.

3 Show that, for all values of t, the point whose position vector is $\mathbf{r} = t\mathbf{i} + (3t + 1)\mathbf{j}$ lies on the straight line whose equation is $y = 3x + 1$.

4 Use the trapezium rule with six strips to evaluate $\int_0^{\frac{\pi}{3}} \tan x \, dx$.

5 a) Find the term in x^5 in the expansion of $(x + \frac{1}{4})^3 (x - \frac{1}{2})^5$.
b) Write $5 + 6x - x^2$ in the form $a - (x - b)^2$. Hence find its maximum value.

6 A geometric progression has a first term of a and a common ratio of $\dfrac{1}{\sqrt{3}}$.

Show that the sum to infinity of the progression is $\frac{1}{2}(3 + \sqrt{3})a$.

7 Eliminate θ from the pair of equations:
$$x = \sin \theta + \cos \theta, \quad y = \tan \theta$$

8 Show that $x + 4\sqrt{x^2 + 1}$ has a minimum value of $\sqrt{15}$.

9 Write down the equation of the tangent to the parabola $y^2 = 4ax$ at the point P $(at^2, 2at)$.

Q is another point on the parabola such that the line PQ passes through the point S $(a, 0)$. Show that the coordinates of Q are $\left(\dfrac{a}{t^2}, -\dfrac{2a}{t}\right)$.

Write down the equation of the tangent at Q and show that the tangents at P and Q meet at T whose y-coordinate is the same as the y-coordinate of the midpoint of PQ.

10 Show that the complex numbers satisfying:
$$|z + 1| = 3|z - 1|$$
represents points lying on a circle in the Argand diagram. Find the Cartesian equation of this circle. What is the radius of the circle? What complex number represents the centre of this circle?

11

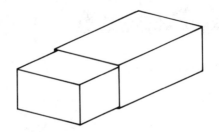

A gift box consists of an outer cover, open at both ends, into which slides an open rectangular box. The length of the box is one-and-a-quarter times its breadth, and the capacity of the box is 20 cm³.

a) If the breadth of the box is x cm, find, in terms of x, the total area of material used.

b) Find the least area of material required to make the gift box. (Assume that the material is of negligible thickness.)

12 Show, by integration, that the volume of a segment of a sphere of radius R is $\frac{1}{3}\pi h^2(3R - h)$, where h is the height of the segment.

Write down an expression for the volume of water in a bowl of radius 10 cm when the depth of water in the bowl is x cm.

An empty hemispherical bowl of radius 10 cm is placed beneath a dripping tap to collect the water. If the tap drips at the rate of 25 cm³ per minute find the rate at which the water level is rising when the depth of water in the bowl is 5 cm. How long has the tap been dripping to give this quantity of water?

Revision Paper 7

1 Show that the sum of the numbers in the rth bracket of the series

$$(1) + (1 + 2) + (1 + 2 + 3) + (1 + 2 + 3 + 4) + \ldots \text{ is } \frac{r}{2}(r + 1). \text{ Hence show}$$

that the sum of n brackets is $\frac{n}{6}(n + 1)(n + 2)$.

2 If $\sec\theta = \frac{4}{3}$ and θ is in the first quadrant, find, in surd form, the value of:

a) $\operatorname{cosec}\theta$,   b) $\tan 2\theta$,   c) $\sec 2\theta$.

3 The numbers $n - 2$, $n + 2$, $3n - 2$ are in GP. Find the two possible values for the common ratio.

4 Express $\dfrac{x}{(1 + x)(1 + 2x)}$ in partial fractions. Hence show that the first three non-zero terms in the expansion of this expression are $x - 3x^2 + 7x^3$.

5 Given that $x > 3y > 0$ and that $x^2 + 9y^2 = 7xy$ prove that:
a) $(x - 3y)^2 = xy$
b) $2 \ln (x - 3y) = \ln x + \ln y$

6 Use Simpson's rule with eight strips to find:

$$\int_2^{10} \sqrt{1 + x^2}\, dx$$

7 Find the coordinates of the two points on the curve $y = x^3 - 3x^2 - 5x + 10$ where the tangents are parallel to the line $y = 4x - 7$. What are the equations of these tangents?

8

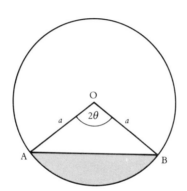

A chord AB divides a circle of radius a into two segments as shown in the diagram. If $A\hat{O}B = 2\theta°$, show that the area of the shaded region is $(\theta - \frac{1}{2} \sin 2\theta)a^2$.

By drawing the graph of $y = \frac{1}{2} \sin 2\theta$, together with another appropriate graph, find the value of θ in radians, correct to two decimal places, if the area of the shaded part is one-quarter the area of the circle.

9 Prove that $\dfrac{x^2 + x - 10}{x - 3}$ can take all values except those between $7 - 2\sqrt{2}$ and $7 + 2\sqrt{2}$. Sketch the graph of $y = \dfrac{x^2 + x - 10}{x - 3}$.

10 Solve the equations:
a) $e^{x^2 - 5x + 8} = e^2$
b) $e^{x^2 + 2x - 8} = 1$
c) $\ln (x^2 - 4x - 11) = 0$

11 Write down an expression for the sum of the squares of the first n natural numbers. Use this formula to find:

a) $1^2 + 2^2 + 3^2 + \ldots + 40^2$

b) $1^2 - 2^2 + 3^2 - 4^2 + \ldots - 40^2$

c) $2^2 + 4^2 + 6^2 + 8^2 + \ldots + 40^2$

d) $20^2 + 21^2 + 22^2 + 23^2 + \ldots + 40^2$

12 Find: a) $\displaystyle\int \frac{dx}{x(x^2 + 1)}$ b) $\displaystyle\int xe^{\frac{x}{2}}\, dx$

Revision Paper 8

1 a) Prove that $\log_b N = \dfrac{\log_{10} N}{\log_{10} b}$. Use this to find the value of $\log_3 8$ correct to three significant figures.

b) Solve the equation $4^{2x + 1} = 2^{x + 4}$.

2 Solve the equation $3^{2x + 1} - 14.3^x - 5 = 0$.

3 Sand falls from an elevator at a constant rate of $\frac{1}{4}$ m^3 per second. It forms a conical heap whose height is equal to the radius of its base. At what rate is the height increasing after one minute?

4 A variable isosceles triangle has a constant perimeter $2s$. If x denotes the length of each of the equal sides find an expression for the area of the triangle, A, in terms of x and s. For what value of x (in terms of s) is the area of the triangle a maximum?

5 When $(1 + x)^3 \sqrt{1 - 3x}$ is expanded in terms of x as far as the term in x^2 the result is $a + bx + cx^2$. Find the values of a, b, c.

6 Show that if the polynomial $f(x)$ is divided by $(x - a)^2$ the remainder is $f(a) + (x - a) f'(a)$.

7 Solve the equation $3(2^x) = 4^{2-x}$ giving the value of x correct to three decimal places.

8 A farmer agreed to give his daughter a certain amount of land each year. He gave her 4 acres the first year, 2 acres the second year, 1 acre the third year, and so on. Assuming that both the farmer and his daughter lived forever how much land would the daughter eventually own?

9 A curve is given by the parametric equations $x = \frac{1}{2}t^2$, $y = t$. Prove that the equation of the tangent at the point where $t = v$ is $vy - x = \frac{1}{2}v^2$.

Write down the equation of the tangent to the curve at the point $(8, 4)$.

Show that the equations of the two tangents from the point $(-4, 1)$ to the curve are $x + 2y + 2 = 0$ and $x - 4y + 8 = 0$. Find the angle between these tangents giving your answer to the nearest degree.

10 The complex number z satisfies the equation:

$$4z\bar{z} - 8z = 13 - 4i$$

where $\bar{z}$ is the complex conjugate of z. Find, in the form $x + yi$, the two possible values of z.

11 P $(2, 4)$, Q $(8, 1)$ and R $(7, -1)$ are the vertices of a triangle. Show that the triangle contains a right angle.
a) Find the area of the triangle PQR.
b) The triangle PQR is reflected in the line PQ to give the triangle $P_1 Q_1 R_1$. Find the coordinates of the vertices of this triangle.
c) The triangle PQR is rotated clockwise about P through $90°$ to give the triangle $P_2 Q_2 R_2$. Find the coordinates of the vertices of this triangle.
d) The triangle PQR is rotated anticlockwise about Q through $180°$ to give the triangle $P_3 Q_3 R_3$. Find the coordinates of the vertices of this triangle.

12 Express $\dfrac{3x^2 + x - 3}{(x - 4)(x + 3)^2}$ in partial fractions. Hence find $\displaystyle\int \dfrac{3x^2 + x - 3}{(x - 4)(x + 3)^2}\,dx$.

Revision Paper 9

1 Solve the equation $\dfrac{x^2}{x + 1} + \dfrac{x + 1}{x^2} = \dfrac{17}{4}$.

2 Solve the simultaneous equations:

$$x^2 + y^2 - 6x = 16, \qquad x^2 + y^2 - 2x - 4y = 8$$

3 If $x = a \sin \alpha \cos \beta$, $y = a \sin \alpha \sin \beta$ and $z = a \cos \alpha$ show that $x^2 + y^2 + z^2 = a^2$.

4 If $f(x) = x \sin x$ show that:

$$\frac{f(x + h) - f(x)}{h} = \sin(x + h) + x \cos\left(x + \frac{h}{2}\right) \cdot \frac{\sin \frac{1}{2}h}{\frac{1}{2}h}$$

Deduce an expression for $f'(x)$, the derivative of $f(x)$, given that $\lim_{\theta \to 0} \frac{\sin \theta}{\theta} = 1$.

5 Solve the equation $\left(\frac{1}{4}\right)^x = 3^{x-4}$.

6 An infinite geometric series with a first term of 3 converges to the sum of 4. Find the first four terms of this series.

7 Show that if x is so small that x^3 and higher powers can be neglected,

$$\sqrt{\frac{1 - x}{1 + x}} = 1 - x + \frac{x^2}{2}$$

By putting $x = \frac{1}{15}$, show that $\sqrt{14} \approx 3\frac{334}{450}$.

8 The equations of two circles are:

$$x^2 + y^2 - 4x + 2y - 20 = 0$$
$$x^2 + y^2 - 14x - 10y + 64 = 0$$

These circles intersect at the point A $(6, a)$. Find the value of a. Use this value of a to find the equation of the tangent to each circle at A. Show that the angle between these tangents is $\tan^{-1} \frac{9}{13}$.

9 When the polynomial $f(x)$ is divided by $x - a$ the remainder is $f(a)$ and when it is divided by $x - b$ the remainder is $f(b)$. Assuming that the degree of $f(x)$ is greater than 2 show that when $f(x)$ is divided by $(x - a)(x - b)$ the remainder is:

$$\frac{f(b) - f(a)}{b - a} x + \frac{bf(a) - af(b)}{b - a}$$

10 Express the complex number $z = \frac{5(1 - i)}{4 + 3i}$ in the form $x + yi$, and calculate its modulus, r, and argument, θ.

Find the moduli and arguments of $-z, \frac{1}{z}, z^2, \bar{z}$ and $\frac{1}{\bar{z}}$.

11 A triangle XYZ has its vertices at the points X $(2, -1, 3)$, Y $(-4, 2, -2)$ and Z $(4, 3, -1)$. Find, in the form $x\mathbf{i} + y\mathbf{j} + z\mathbf{k}$, the vectors $\overrightarrow{YX}$, $\overrightarrow{YZ}$ and $\overrightarrow{ZX}$, and hence find the lengths of the sides of the triangle. Find the scalar product of $\overrightarrow{XY}$ and $\overrightarrow{ZY}$ and use it to calculate the angle XYZ, giving your answer correct to the nearest tenth of a degree.

12

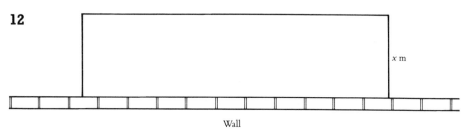

Wall

A farmer has 100 m of fencing. She wishes to enclose a rectangular area of ground using an existing wall as one side. If the rectangle is x m wide, find an expression for the area enclosed (A m^2) in terms of x. Hence find the largest area that can be enclosed in this way by 100 m of fencing.

Is it possible to erect the fencing in a different way so that an even greater area is enclosed?

Revision Paper 10

1 Solve the simultaneous equations:

$$4x^2 - 2xy - y^2 = 4, \quad 2x - 3y + 8 = 0$$

2 Solve the equation $6 \cos 2x + 5 \cos x - 4 = 0$ for $0° \leqslant x \leqslant 360°$.

3 Show that the general term of the series $2 \times 2 + 5 \times 3 + 8 \times 4 + \ldots$ is given by $T_r = (3r - 1)(r + 1)$.

Hence show that the sum of n terms of this series is $\dfrac{n}{2}(2n^2 + 5n + 1)$.

4 Find a formula for $\tan (A + B + C)$ in terms of $\tan A$, $\tan B$ and $\tan C$.

Deduce that if A, B and C are the angles of a triangle then
$\tan \frac{1}{2} A \tan \frac{1}{2} B + \tan \frac{1}{2} B \tan \frac{1}{2} C + \tan \frac{1}{2} C \tan \frac{1}{2} A = 1$.

5 If the expansion of $\dfrac{1}{(1 + x)(1 + 2x)(1 + 3x)}$ as far as the term in x^3 is $1 - 6x + ax^2 + bx^3$ find the values of a and b.

6 The first term of a geometric progression is $\sqrt{2}$ and the common ratio is $-\dfrac{1}{\sqrt{2}}$. Show that the sum to infinity of the progression is $2(\sqrt{2} - 1)$.

7 Prove that $\dfrac{(x-2)(x-1)}{(x-3)}$ can take all values except those between $3 - 2\sqrt{2}$ and $3 + 2\sqrt{2}$. Sketch the graph of $y = \dfrac{(x-2)(x-1)}{(x-3)}$.

8 Solve the simultaneous equations:

$$3x^2 + 3xy = 22, \qquad 2xy + 3y^2 = 15$$

9 Prove that $\displaystyle\sum_{r=1}^{n} r = \dfrac{n}{2}(n+1)$.

Show that $(r+1)^3 - r^3 \equiv 3r^2 + 3r + 1$.

Use this identity, for values of r from 1 to n, to find a formula for the sum of the squares of the first n natural numbers.

10 a) P is a variable point $(2t^2 + 1, t - 1)$. Show that the locus of P has equation $2y^2 + 4y - x + 3 = 0$.

b) P is a variable point $(a \sec \theta, b \tan \theta)$. Show that the locus of P as θ varies has equation $\dfrac{x^2}{a^2} - \dfrac{y^2}{b^2} = 1$.

11

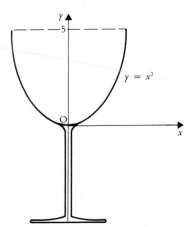

The bowl of a wine glass is formed by rotating that part of the curve $y = x^3$, for values of y from 0 to 5, through four right angles about the y-axis. Show that the amount of wine in the glass when the depth is y is $\frac{3}{5}\pi \sqrt[3]{y^5}$ cm^3. Find the capacity of the glass if it is 5 cm deep.

Wine is poured into a similar wine glass at a steady rate of 20 cm^3/second. At what rate is the level rising when the depth is 2 cm?.

12 a) Show that $\displaystyle\int_0^{\frac{\pi}{6}} \sin 2x \sqrt{\sin x}\, dx = \dfrac{\sqrt{2}}{10}$.

b) Find: **i)** $\displaystyle\int \dfrac{dx}{1 + 4x^2}$ **ii)** $\displaystyle\int \dfrac{(2x+3)}{x^2 + 1}\, dx$

Part 3: Past A-level Questions, Papers 1–15

Paper 1

1 Solve the equation

$$\tan^2 \theta + 3 \sec \theta = 0,$$

giving all solutions in degrees in the interval $0° \leqslant \theta \leqslant 360°$. *(6 marks)*
(AEB, 1988)

2 Given that k is real, find the set of values of k for which the roots of the quadratic equation

$$(1 + 2k)x^2 - 10x + (k - 2) = 0$$

a) are real, *(3 marks)*
b) have a sum which is greater than 5. *(3 marks)*
(AEB, 1988)

3 a) Show that $\log_9(xy^2) = \frac{1}{2}\log_3 x + \log_3 y$. *(3 marks)*
   Hence, or otherwise, solve the simultaneous equations

$$\log_9(xy^2) = \frac{1}{2}$$
$$(\log_3 x)(\log_3 y) = -3.$$ *(6 marks)*

b) An arithmetic progression has first term $\ln 2$ and common difference $\ln 4$. Show that the sum S_n of the first n terms is $n^2 \ln 2$.

Find the least value of n for which S_n is greater than fifty times the nth term. *(7 marks)*
(AEB, 1988)

4 The nth term of a geometric progression is λ times the first term. Prove that the common ratio, r, of the progression is given by $\log r = \dfrac{\log \lambda}{n - 1}$, where the base of the logarithms is optional.

At the end of 1970, the population of Newtown was 46 650 and by the end of 1975 it had risen to 54 200. On the assumption that the population increased annually in a geometric progression between 1970 and 1975 and continued likewise afterwards, find

a) the common ratio of the progression, giving the value correct to 3 significant figures,
b) the population at the end of 1985. *(7 marks)*
(SU)

5 The functions f, g and h are defined by

$$f : x \mapsto \ln x, \qquad (x \in \mathbb{R}^+),$$
$$g : x \mapsto \frac{1}{x}, \qquad (x \in \mathbb{R}^+),$$
$$h : x \mapsto x^2, \qquad (x \in \mathbb{R}).$$

a) Give definitions of each of the functions fg and f^{-1}, and state, in each of the following cases, a relationship between the graphs of
 i) f and fg,
 ii) f and f^{-1}. (4 marks)

b)

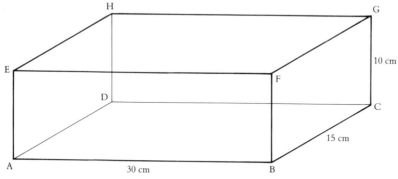

The diagram shows a sketch of the graph of $y = hf(x)$. State how the sketch shows that hf is not one–one, and prove that, if α and β, where $0 < \alpha < \beta$, are such that $hf(\alpha) = hf(\beta)$, then $\alpha = g(\beta)$. (5 marks)

c) The function ϕ is defined by

$$\phi : x \longmapsto hf(x), \qquad (0 < x \leqslant 1).$$

Sketch the graph of ϕ^{-1}, and give an explicit expression in terms of x for $\phi^{-1}(x)$. (3 marks)

(C)

6

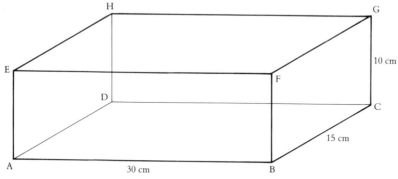

The diagram shows a rectangular solid ABCDEFGH, with AB = 30 cm, BC = 15 cm and CG = 10 cm. Calculate

a) the angle between DG and BG, giving your answer correct to the nearest 0.1°, (5 marks)

b) the perpendicular distance of C from BD, giving your answer correct to three significant figures, (4 marks)

c) the angle between the planes DGB and DCB, giving your answer correct to the nearest 0.1°. (3 marks)

(C)

7 Given the curve $y = 2 + 2x + x^2 + \frac{1}{3}x^3$, prove that

a) the gradient at any point is always positive;

b) there is a point of inflexion at $(-1, \frac{2}{3})$;

c) the gradient at $(-1, \frac{2}{3})$ is 1.

Sketch the curve, clearly indicating the features proved.

(A formal graph will not be accepted; a sketch for all values of x is required.)

(SU)

8 (In this question the formulae $\sin 3x \equiv 3 \sin x - 4 \sin^3 x$ and $\cos 3x \equiv 4 \cos^3 x - 3 \cos x$ may be used.)

Find the coordinates of all the stationary points on the curve $y = \frac{1}{2} + \sin 3x - 3 \sin x$ in the range $0 \leqslant x \leqslant 2\pi$.

Classify, with reasons, each point as a maximum point, a minimum point or a point of inflexion.

Determine all the points in the same range where the curve crosses the x-axis. Sketch the graph of $y = \frac{1}{2} + \sin 3x - 3 \sin x$ in the range $0 \leqslant x \leqslant 2\pi$.

(SU)

9 A function $y = f(x)$ is tabulated for various values of x as shown below:

x	1.0	1.2	1.4	1.6	1.8
y	3.70	3.82	4.15	4.51	5.07

Estimate

a) the value of y at $x = 1.15$,

b) the value of x for which $y = 4.40$,

c) $\displaystyle\int_1^{1.8} y \, dx$, using Simpson's rule.

(O&C)

10 Find values of a and b in order that

$$x^2 - 4x + 13 = (x - a)^2 + b, \quad \text{for all values of } x.$$

Hence evaluate

$$\int_2^3 \frac{dx}{x^2 - 4x + 13}$$

correct to 3 decimal places.

(O&C/MEI)

Paper 2

1 a) In the expansion of
$$(x + a)^5 (x - 2a)^3,$$
the coefficients of x^6 and x^7 are equal and non-zero.

Evaluate a.

b) The roots of the quadratic equation
$$x^2 - px + q = 0$$
are real and one is the cube of the other. Express p in terms of q. (NI)

2 The rectangle ABCD is the horizontal base of a pyramid, and the vertex V of the pyramid is vertically above the centre of the base. The length of AB is 6 cm, the length of BC is 12 cm and the length of VA is 20 cm. The points X on VB and Y on BC are such that AX and YX are both perpendicular to VB.

a) Show that the length of BX is 0.9 cm, and calculate the length of BY.
(6 marks)

b) Calculate, to the nearest degree, the angle between the planes VAB and VBC.
(6 marks)
(C)

3 A circle whose centre is in the first quadrant touches the x-axis at $(3, 0)$ and also touches the line $4y = 3x + 36$.

a) Determine the equation of the circle and the coordinates of its point of contact with this line.

b) There are two parallel tangents to the circle which are at right angles to the line $4y = 3x + 36$. Find the coordinates of the points where these two tangents meet the line. (NI)

4 The function f is defined by
$$f : x \mapsto 4x^3 + 3, \qquad (x \in \mathbb{R}).$$
Give the corresponding definition of f^{-1}.
State the relationship between the graphs of f and f^{-1}. (C)

5 Relative to the origin O, the position vectors of the points A, B and C are $\mathbf{j} - 4\mathbf{k}$, $6\mathbf{i} - 5\mathbf{j} - \mathbf{k}$ and $4\mathbf{i} + 7\mathbf{j} - 9\mathbf{k}$ respectively, the unit of length being the metre.

a) Show that, for all values of the scalar parameter t, the point P with position vector $2t\mathbf{i} + (1 - 2t)\mathbf{j} + (t - 4)\mathbf{k}$ lies on the straight line passing through A and B.
(2 marks)

b) Use the scalar product $\overrightarrow{AB} \cdot \overrightarrow{CP}$ to determine the value of t for which CP is perpendicular to AB.
(4 marks)

c) Hence find the shortest distance from C to AB.
(3 marks)
(AEB, 1986)

103

6

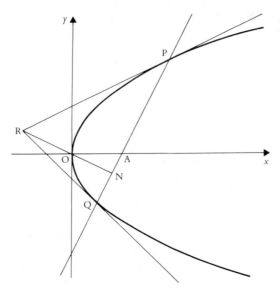

A parabola is defined by the parametric equations

$$x = t^2, \qquad y = 2t.$$

Find the equations of the tangents at the points P $(p^2, 2p)$ and Q $(q^2, 2q)$. Show that these tangents intersect at the point R $(pq, p + q)$.

Show that the equation of the line PQ is

$$(p + q)y = 2x + 2pq.$$

The points P and Q move on the parabola in such a way that pq remains constant and equal to -2. Prove that

a) the line PQ always passes through the point A $(2, 0)$.

b) the line through R and the origin O is always perpendicular to PQ.

The line RO meets PQ at the point N, as shown in the diagram. Show that N moves on a fixed circle, and state (or find) the coordinates of the centre, and the radius, of this circle. (JMB)

7 A rectangular box with a lid is to be made of uniformly thin material so that its length is three times its breadth.

a) Express each of the volume and surface area of the box as a function of the breadth and depth.

b) If the volume of the box is to be 36 m³, show that the minimum possible area of the surface is 72 m².

c) If, on the other hand, the surface area of the box is to be 72 m², show that the maximum possible volume is 36 m³. (NI)

8 a) Define the inverse tangent function $\tan^{-1} x$, and show that

$$\frac{d}{dx}(\tan^{-1} x) = \frac{1}{1+x^2}.$$

Use Maclaurin's theorem to show that

$$\tan^{-1} x = x - \frac{x^3}{3} + \frac{x^5}{5} - \cdots$$

b) i) Express $\tan 4\theta$ as a function of $\tan \theta$.

ii) If $\theta = \tan^{-1} \frac{1}{5}$ and $\phi = \tan^{-1} \frac{1}{239}$, show that $4\theta - \phi = \frac{\pi}{4}$.   **(NI)**

9 a) Show that $\dfrac{d}{dx}\{e^{2x}(\cos 3x + k \sin 3x)\}$, where k is a constant, may be expressed in the form $e^{2x}(A \cos 3x + B \sin 3x)$, and find A and B in terms of k.

By choosing suitable values for k, or otherwise, find $\displaystyle\int e^{2x} \cos 3x\, dx$ and $\displaystyle\int e^{2x} \sin 3x\, dx$.

b) Evaluate $\displaystyle\int_0^{\frac{\pi}{3}} \cos^3\left(\tfrac{1}{2}x\right) dx$.   **(O&C)**

10 a) Find $\displaystyle\int_0^{\frac{\pi}{4}} \frac{\sin \theta}{\cos^4 \theta}\, d\theta$ in terms of $\sqrt{2}$.

b) Express $\dfrac{4x+2}{(x-1)(2x^2+1)}$ in partial fractions, and hence find

$\displaystyle\int_2^3 \frac{4x+2}{(x-1)(2x^2+1)}\, dx$, expressing your answer in the form $\ln(a/b)$ where a and b are integers.

(16 marks)
(O&C/MEI)

Paper 3

1 a) If $\sin 3\theta = \sin 2\theta$ and $\sin \theta \neq 0$, show that

$$4\cos^2\theta - 2\cos\theta - 1 = 0.$$

b) Deduce that

$$\cos\frac{\pi}{5} = \frac{1}{4}(1 + \sqrt{5}), \text{ and}$$

$$\cos\frac{3\pi}{5} = -\frac{1}{4}(-1 + \sqrt{5}).$$

    (NI)

2 a) For which positive integer b is $(x + b)$ a factor of $x^3 + x^2 - x + 5b$?

b) Prove that, for all values of θ,

$$-29 \leqslant 20 \sin \theta + 21 \cos \theta \leqslant 29. \qquad \text{(NI)}$$

3 It is given that one root of the equation

$$\cos x - 6 \sin x = x^2$$

is sufficiently small for powers of x above the second to be neglected. Use approximations for $\cos x$ and $\sin x$ to obtain an estimate for this root to two decimal places. *(4 marks)*

(JMB)

4 P and Q are points on a circle of radius a, and the minor arc PQ subtends an angle 2θ radians at its centre O.

a) Write down the area A of the sector POQ. *(3 marks)*

b) Show that the area B enclosed by the arc PQ and the tangents to the circle at P and Q is given by

$$B = a^2 (\tan \theta - \theta). \qquad \text{(4 marks)}$$

c) Given that $\theta = \dfrac{3\pi}{8}$ radians is an approximate solution to the identity $A = B$, use the Newton–Raphson method (applied once) to obtain a better approximation. *(5 marks)*

(NI)

5 Three points A $(1, 1)$, B $(7, 4)$, C $(8, 2)$ are given. Show that the triangle ABC is right angled.

a) If the triangle is rotated clockwise about B through $90°$ into the position A'B'C', what are the coordinates of A', B', C'?

b) If, instead, the triangle is reflected in the line AB into the position A''B''C'', find the coordinates of A'', B'', C''. *(O&C)*

6 Let $f(x) = \cos^{-1} x$, where $-1 < x < 1$ and $0 < f(x) < \pi$. Show that

$$(1 - x^2)f''(x) - xf'(x) = 0 \qquad \text{(4 marks)}$$

By differentiating this result three times, or otherwise, show that $f^{(5)}(0) = -9$, and find the first four non-zero terms in the expansion of $f(x)$ in ascending powers of x. *(8 marks)*

(C)

7

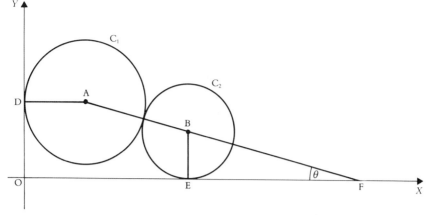

The figure shows two perpendicular axes OX and OY. Also shown are two circles C_1 and C_2 which touch each other and which both lie in the first quadrant (i.e. above OX and to the right of OY).

C_1 has radius 4 and touches OY at D; C_2 has radius 3 and touches OX at E. The line AB, joining the centres of C_1 and C_2, meets OX at F and $\widehat{BFO} = \theta$.

a) Find expressions for OD and OE in terms of θ and show that

$$DE^2 = 74 + 42 \sin \theta + 56 \cos \theta.$$

Hence express DE^2 in the form

$$74 + r \cos (\theta - \alpha)$$

where the values of r and α are to be found. *(6 marks)*

b) By considering the extreme positions in which
   i)   both circles touch OX
   and ii) both circles touch OY,
   show that, correct to 1 decimal place,

$$8.2° \leqslant \theta \leqslant 98.2°.$$ *(4 marks)*

c) Find the greatest and least possible lengths of DE and state the corresponding values of θ. *(5 marks)*

(W)

8 Write down the expansions of $\ln (1 + x)$ and e^{-x} as far as the term in x^4.

Given that $f(x) \equiv 1 + x - \ln (1 + x)$, prove that the expansion of $g(x) \equiv (1 + x) f(x)$ as far as the term in x^4 is $1 + x + \frac{1}{2}x^2 + \frac{1}{6}x^3 - \frac{1}{12}x^4$.

Given that $G(x) \equiv (1 - x) F(x)$ where $F(x) \equiv 1 - x - \ln (1 - x)$, write down the expansion of $G(x)$ as far as the term in x^4.

Prove that the expansion of $h(x) \equiv e^{-x} g(x)$ as far as the term in x^4 is $1 - \frac{1}{8}x^4$.

(SU)

9 a) The finite region bounded by the x-axis, the curve $y = e^{-x}$, and the lines $x = \pm a$ is denoted by R. Find, in terms of a, the volume of the solid generated when R is rotated through one revolution about the x-axis.

(*3 marks*)

b) i) By using a suitable substitution, or otherwise, evaluate

$$\int_0^1 x(1-x)^9 \, dx.$$

(*4 marks*)

ii) Find $\displaystyle\int 2x \tan^{-1} x \, dx.$

(*5 marks*)

(C)

10 Indicate on an Argand diagram the set S of complex numbers z which satisfy the inequality

$$|z - (8 + 6i)| \leqslant 3.$$

Find the least value of $|z|$ for $z \in S$. Calculate correct to three significant figures the corresponding value, θ, of $\arg z$, where $-\pi < \theta \leqslant \pi$.

Mark on your diagram the least value, α, of $\arg z$ for $z \in S$.

(JMB)

Paper 4

1 a) Factorise $4x^2 - 8xy + 3y^2$.

Solve the simultaneous equations $4x^2 - 8xy + 3y^2 = 8$,

$$2x - 3y = 4.$$

b) Solve the equation $3(9^x) - 28(3^x) + 9 = 0$.

(*6 marks*)

(SU)

2 Given that $f(x) \equiv x^3 - 14x + 8$ and that $f(x) = 0$ has a negative integral root, find all the values of x (to 2 decimal places where suitable) for which $f(x) = 0$.

Sketch the curve $y = f(x)$.

State the ranges of values of x for which $f(x) < 0$, using decimal values (to 2 decimal places) where necessary.

(*6 marks*)

(SU)

3 Given that $3 \cos \theta + 4 \sin \theta \equiv R \cos (\theta - \alpha)$, where $R > 0$ and $0 \leqslant \alpha \leqslant \pi/2$, state the value of R and the value of $\tan \alpha$. (2 marks)

For each of the following equations, solve for θ in the interval $0 \leqslant \theta \leqslant 2\pi$ and give your answers in radians correct to one decimal place.
a) $3 \cos \theta + 4 \sin \theta = 2$, (3 marks)
b) $3 \cos 2\theta + 4 \sin 2\theta = 5 \cos \theta$, (6 marks)

The curve with equation $y = \dfrac{10}{3 \cos x + 4 \sin x + 7}$, between $x = -\pi$ and $x = \pi$, cuts the y-axis at A, has a maximum point at B and a minimum point at C. Find the coordinates of A, B and C. (5 marks)

(AEB, 1988)

4 A pyramid VABCD has a horizontal square base ABCD, of side $6a$. The vertex V of the pyramid is at a height $4a$, vertically above the centre of the base. Calculate, in degrees, to one decimal place, the acute angle between

a) the edge VA and the horizontal, (2 marks)
b) the plane VAB and the horizontal, (1 mark)
c) the planes VBA and VBC. (5 marks)

(AEB, 1988)

5 a) Find the complex numbers z_1 and z_2 which satisfy the simultaneous equations

$$2z_1 + z_2 = 7 + 2i,$$

$$z_1 + iz_2 = 6 + 6i.$$

Give your answers in the form $x + iy$ (where x and y are real).
b) If $z = x + iy$, where x and y are real, find an expression for the value of a when $\dfrac{z - 8i}{z - 6}$ is given in the form $a + ib$ (a, b real).

Given that $z = x + iy$, where x and y are real, and that $\dfrac{z - 8i}{z - 6}$ is purely imaginary, prove that the locus of the point representing z in the Argand diagram is a circle, and find its centre and radius. (SU)

6 a) If $x \geqslant 0$, sketch the curve whose equation is

$$y = \frac{3\sqrt{x}}{1 + 2x^2}.$$

b) If the region bounded by this curve, the x-axis, and the ordinate $x = h$ (where h is large), is rotated about the x-axis through an angle of 2π radians, show that the volume of rotation is almost $\dfrac{9\pi}{4}$. (NI)

7 Prove that the equation of the chord joining the points P $(cp, c/p)$ and Q $(cq, c/q)$ on the rectangular hyperbola $xy = c^2$ is
$$pqy + x = c(p + q).$$
Deduce, or find otherwise, the equation of the tangent at P to the hyperbola.

Find the coordinates of T, the point of intersection of the tangents at P and Q.

For all chords PQ passing through a fixed point R (h, k), prove that T always lies on a fixed line l. The line l meets the x-axis at M and the line OR through the origin O at N; prove that the triangle OMN is isosceles. Find also the locus of the mid-point of PQ.   (O&C)

8 a) When an employee retires from a particular firm, a lump-sum payment is made depending on the number of years of full-time employment with the firm. At the end of the first year of employment £300 is set aside towards this lump-sum. At the end of the second year another £300 is added and interest at 6% is added to the first year's £300. At the end of the third year another £300 is added and interest at 6% is added to the total sum of money which had accrued by the end of the second year. This process continues for the number of completed years of employment. By forming and summing a series, prove that an employee would receive $5000[(1.06)^n - 1]$ pounds as the lump-sum payment for n years of employment with this firm.   *(4 marks)*

Hence find the least number of years required for a lump-sum to exceed £15 000.   *(3 marks)*

b) Another employee retired with savings of just under £10 000. He records the value of these savings (S pounds) at the end of t years after retirement. The results are shown in the following table.

t	3	5	8	12
S	5100	3100	1700	700

The value of his savings is believed to follow a law of the form $S = A(B^t)$, where A and B are constants. By drawing an appropriate linear graph, verify that this belief is approximately valid and use your graph to estimate values for A and B.   *(9 marks)*

(AEB, 1988)

9 a) Find the coordinates of the turning points on the curve with equation
$$y = \frac{x^2}{1 + x^4}.$$
Sketch the curve.   *(7 marks)*

b) A curve has parametric equations $x = 5a \sec \theta$, $y = 3a \tan \theta$, where $-\frac{1}{2}\pi < \theta < \frac{1}{2}\pi$ and a is a positive constant. Find the coordinates of the point on the curve at which the normal is parallel to the line $y = x$.   *(5 marks)*

(C)

10 a) Given that $y = \tan^{-1} x$, prove that $\dfrac{dy}{dx} = \dfrac{1}{1 + x^2}$.

Using integration by parts, find $\displaystyle\int \tan^{-1} x \, dx$.

By expressing $x^2 + 6x + 10$ in the form $(x + a)^2 + b$, evaluate

$$\int_{-3}^{-2} \frac{dx}{x^2 + 6x + 10}.$$

b) Sketch the curve $x^2 = ay$.

A portion of this curve is rotated about the vertical axis Oy to form an open bowl with a flat circular base.

If the radius of the base is r and the radius of the top of the bowl is $3r$, prove that the depth of the bowl is $\dfrac{8r^2}{a}$.

Given that the volume of the bowl is $\dfrac{2}{5}\pi a^3$, find the depth of the bowl in terms of a.

(SU)

Paper 5

1 A triangle ABC has angles $A = 40°$, $B = 60°$, $C = 80°$. Write down an expression for the ratio of the sides $a : b : c$.

Calculate, to 3 places of decimals, what proportion the side a is of the total perimeter.

(O&C/SMP)

2 a) i) Find an expression for the sum to infinity of the geometric progression $4x + 4x^2 + 4x^3 + \ldots$.

ii) Use the binomial theorem to expand $\dfrac{1 + 3x}{1 - x}$ as far as, and including, the term in x^5.

iii) Show clearly how the sum of the geometric progression in i) is related to $\dfrac{1 + 3x}{1 - x}$.

b) Prove that the coefficient of x^r, where $r > 1$, in the expansion of $\dfrac{1 + 3x}{1 - 2x}$ is $5(2^{r-1})$.

(6 marks)

(SU)

3 The triangle ABC has vertex A (7, 8) and the midpoint of BC is D (4, 2). The perpendicular bisector of BC cuts AB at P (3, 4).

a) Find the gradient of PD and hence find the Cartesian equation of the straight line passing through B and C. *(3 marks)*

b) Find the Cartesian equation of the straight line passing through A and P and hence show that the coordinates of B are (−2, −1). *(4 marks)*

A circle is drawn with AB as diameter.

c) Determine the Cartesian equation of the circle and find the coordinates of the point Q where the circle crosses the line BC between B and C. Show that BQ : QC = 9 : 1 *(9 marks)*

(AEB, 1988)

4 The points A, B and C have position vectors **a**, **b** and **c** respectively. Find the position vector of the point P on AB such that $\overrightarrow{PB} = 3\,\overrightarrow{AP}$. Show that if $\overrightarrow{PC}$ is perpendicular to $\overrightarrow{AB}$,

$$4\mathbf{c}.(\mathbf{b} - \mathbf{a}) = (3\mathbf{a} + \mathbf{b}).(\mathbf{b} - \mathbf{a})$$ *(6 marks)*

(O&C/MEI)

5 A manufacturer produces solid bodies, as shown in the diagram, consisting of a central cylinder, radius r and length h, with two hemispheres, each of radius r, one at each end of the cylinder.

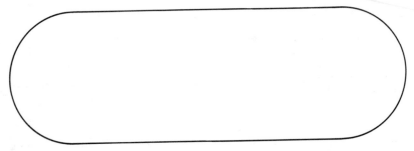

Given that the volume of the body is 36π cubic units, prove that

a) $h = \dfrac{108 - 4r^3}{3r^2}$;

b) the total surface area is $\dfrac{4}{3}\pi r^2 + \dfrac{72\pi}{r}$ square units;

c) the least total surface area is 36π square units.

The manufacturer now decides to modify the design so that the solid body will still have a volume of 36π cubic units, but will consist of a cylinder with just one hemisphere attached. Show that the least surface area is increased to almost 39π square units.

(SU)

6 a) A curve is described parametrically by the equations

$$x = \frac{1+t}{t}, \quad y = \frac{1+t^3}{t^2}.$$

Find the equation of the normal to the curve at the point where $t = 2$.

(6 marks)

b) The points A and B lie on the curve with equation $y = \dfrac{1}{(x+1)^2}$. The x-coordinates of the points A and B are 1 and $1+h$ respectively. Show that the gradient of the line AB is $-\dfrac{(h+4)}{4(2+h)^2}$.

Deduce the gradient of the tangent to the curve at A. (6 marks)

c) Find the gradient at the point $(2, 3)$ on the curve with equation

$$3x^2 + 6xy - 2y^3 + 6 = 0$$

(4 marks)

(AEB, 1988)

7 Given that $y = \ln\left[\frac{1}{2}(1 + e^{-x})\right]$, show that $\dfrac{dy}{dx} = \frac{1}{2}e^{-y} - 1$. (4 marks)

By repeated differentiation of this result, or otherwise, show that the series expansion of y in ascending powers of x, up to and including the term in x^4, is

$$-\tfrac{1}{2}x + \tfrac{1}{8}x^2 + kx^4,$$

where the numerical value of k is to be determined. (8 marks)

(C)

8 A curve is given in terms of the parameter t by the equations

$$x = a\cos^2 t, \quad y = a\sin^3 t, \quad 0 < t < \frac{\pi}{2},$$

where a is a positive constant.

Find and simplify an expression for $\dfrac{dy}{dx}$ in terms of t. (3 marks)

The normal to the curve at the point where $t = \pi/6$ cuts the y-axis at the point N. Find the distance ON in terms of a, where O is the origin.

(4 marks)

(AEB, 1988)

9 a) Expand $\ln (1 - x)(1 - 2x)$ as a series in ascending powers of x as far as, and including, the term in x^5.

Hence find an approximate value (correct to 4 decimal places) of

$$\int_0^{0.1} \ln (1 - 3x + 2x^2) \, dx.$$

b) Given that the expansions of $\dfrac{1 + ax}{1 + bx}$ and e^{2x} are identical as far as the term in x^2, find the values of a and b.

With these values of a and b, prove that

$$\frac{1 + ax}{1 + bx} - e^{2x} = \frac{2}{3}x^3 + \dots$$

(SU)

10 a) By using the method of partial fractions, or otherwise, show that

$$\int_1^2 \frac{x + 3}{x(x + 2)} \, dx = \tfrac{1}{2} \log_e 6.$$

b) By using the substitution $t = \tan \dfrac{\theta}{2}$, or otherwise, show that

$$\int_0^{\frac{\pi}{3}} \frac{d\theta}{1 + \sin \theta} = \sqrt{3} - 1.$$

c) By using integration by parts, or otherwise, obtain

$$\int x^3 (x^2 - 3)^{3/2} \, dx.$$

(NI)

Paper 6

1 Write the expression $2x^2 + 4x + 5$ in the form $a(x + b)^2 + c$, where a, b, c are numbers to be found.

Use your answer to **write down** the coordinates of the minimum point on the graph of $y = 2x^2 + 4x + 5$.

(O&C/SMP)

2 Find the values of x for which

$$x^2 - x + 4 < |4x - 2|.$$

(7 marks)
(O&C/MEI)

3 By first comparing the coefficients of x^3 in the identity

$$x^4 - 5x^3 - 19x^2 + 29x + 42 \equiv (x^2 + ax + 3)(x^2 + bx + 14)$$

and then comparing the coefficients of x, find the values of a and b.

Solve the equation $x^4 - 5x^3 - 19x^2 + 29x + 42 = 0$.

What are the solutions of the equation

$$y^4 + 5y^3 - 19y^2 - 29y + 42 = 0?$$

(SU)

4 A geometric progression has first term 1 and common ratio $\frac{1}{2} \sin 2\theta$.
  a) Find the sum of the first 10 terms in the case when $\theta = \pi/4$, giving your answer to 3 decimal places.
  (2 marks)
  b) Given that the sum to infinity is 4/3, find the general solution for θ in radians.
  (4 marks)
(AEB, 1988)

5 Show that the equation $x^3 - 5x + 3 = 0$ has a root between $x = 0$ and $x = 1$. Using the iteration formula $x_{n+1} = \dfrac{3 + x_n^3}{5}$, find the value of the root to 2 decimal places.

(7 marks)
(O&C/MEI)

6 The base K of a vertical tower is due north of, and at the same horizontal level as, a point A. The point B is due east of the tower but is at a level 8 metres higher than A and K. The angles of elevation of the top of the tower from A and B are 60° and 45° respectively, while the distance from A to B is 40 metres.

Show that the height of the tower is just under 40 metres, and calculate the angle at which the plane KAB is inclined to the horizontal.

(NI)

115

7

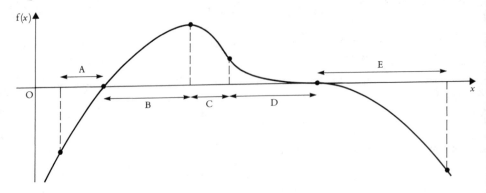

State the signs of f(x), f′(x) and f″(x) inside each of the intervals A, B, C, D, E in the diagram. Set your answers out in the form of a table.

Interval ...		A	B	C	D	E
Signs of	f(x)					
	f′(x)					
	f″(x)					

(O&C/SMP)

8 Sketch the graph of $y = \dfrac{x}{\sqrt{4 + x^2}}$, showing any asymptotes.

Find the area contained between the curve and the x-axis for $0 \leqslant x \leqslant 2$.
Find also the volume generated when this area is rotated through 2π radians about the x-axis.

(16 marks)
(O&C/MEI)

9 A and B are two points on level ground and B is *a* metres due east of A.

A tower of height *h*, standing on the same level ground, is in a direction NθE of A and NϕW of B. The angle of elevation of the top of the tower from A is α and from B is β.

Prove that:

a) $h \sin (\theta + \phi) = a \cos \phi \tan \alpha$;

b) $\phi = \cos^{-1} \dfrac{\cos \theta \tan \beta}{\tan \alpha}$;

c) $h^2 (\cot^2\alpha - \cot^2\beta) - 2ah \cot \alpha \sin \theta + a^2 = 0$.

(SU)

10 Given that $\mathbf{p} = t^2\mathbf{i} + (2t + 1)\mathbf{j} + \mathbf{k}$ and $\mathbf{q} = (t - 1)\mathbf{i} + 3t\mathbf{j} - (t^2 + 3t)\mathbf{k}$ where t is a scalar variable, determine

a) the values of t for which $\mathbf{p}$ and $\mathbf{q}$ are perpendicular, *(3 marks)*

b) the angle between the vectors $\mathbf{p}$ and $\mathbf{q}$ when $t = 1$, giving your answer to the nearest 0.1°. *(4 marks)*

(AEB, 1988)

Paper 7

1 a) Solve the equation $\log (x^2 - 10) - \log x = 2 \log 3$, where each of the logarithms has the same base.

b) Solve the inequality $\dfrac{x - 1}{x - 2} > 3$.

c) Solve the equation $|1 + 3x| = 1$. *(SU)*

2 a) Expand

$$(1 + x)^{\frac{1}{5}} \quad \text{and} \quad \frac{5 + 3x}{5 + 2x}$$

in ascending powers of x, up to and including the term involving x^3. Verify that the first three corresponding terms in each expression coincide.

b) Deduce that $\sqrt[5]{\dfrac{21}{20}} \approx \dfrac{103}{102}$. *(NI)*

3

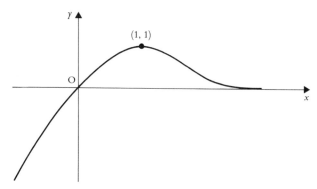

(1, 1)

The diagram shows the graph of $y = f(x)$. The curve passes through the origin, and has a maximum point at (1, 1). Sketch, on separate diagrams, the graphs of

a) $y = f(x) + 2$,   b) $y = f(x + 2)$,   c) $y = f(2x)$,

giving the coordinates of the maximum point in each case. *(6 marks)*

(C)

4 The coordinates of the points A and C are $(-1, 3)$ and $(4, 2)$.

Show that the equation of the circle having AC as diameter is

$$x^2 + y^2 - 3x - 5y + 2 = 0.$$

Given that ABCD is a square, and B is the point with coordinates $(1, 0)$, find the coordinates of D.

(SU)

5 a) Write $\sin X + \sin Y$ in product (factor) form.

Solve the equation $\sin (x + 70°) + \sin (x - 50°) = 0.5$, giving all solutions between $0°$ and $360°$.

b)

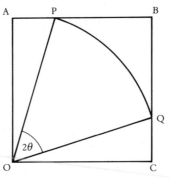

In the diagram, OABC is a square with OA $= a$.

The circular sector OPQ, with angle POQ $= 2\theta$ radians, has area $\frac{1}{2}a^2$.

Prove that i) the radius of the sector is $\dfrac{a}{\cos \left(\frac{\pi}{4} - \theta\right)}$;

ii) $\sin 2\theta + 1 = 4\theta$.

By plotting the graphs of $y = \sin 2\theta + 1$ and $y = 4\theta$, deduce the value of θ. Use a scale of 1 cm $= 0.1$ on the θ axis and 1 cm $= 0.5$ on the y axis.

(SU)

6 Show that

$$\frac{(x-1)^2}{3-x} = \frac{4}{3-x} - (x + 1).$$

For the curve $y = \dfrac{(x-1)^2}{3-x}$, find the coordinates of the stationary points, distinguishing between maximum and minimum values.

Sketch the curve.

Find the area under the curve between $x = 0$ and $x = 1$. *(16 marks)*

(O&C/MEI)

118

7 A sector S of a circle, of radius R, whose angle at the centre of the circle is ϕ radians, is rolled up to form the curved surface of a right cone standing on a circular base. The semi-vertical angle of this cone is θ radians. Express ϕ in terms of $\sin \theta$ and show that the volume V of the cone is given by

$$3V = \pi R^3 \sin^2 \theta \cos \theta. \qquad \text{(5 marks)}$$

If R is constant and θ varies, find the positive value of $\tan \theta$ for which $\dfrac{dV}{d\theta} = 0$. Show further that when this value of $\tan \theta$ is taken, the maximum value of V is obtained. *(6 marks)*

Hence show that the maximum value of V is $\dfrac{2\pi R^3 \sqrt{3}}{27}$ and find, in terms of R, the area of the sector S in this case. *(5 marks)*

(AEB, 1988)

8 [Use 2 mm graph paper for this question.]

The variables x and y are known to satisfy an equation of the form $y = ab^x$, where a and b are constants. For five different values of x, corresponding approximate values of y were obtained experimentally. The results are given in the following table.

x	2.0	2.5	3.0	3.5	4.0
y	11.3	18.0	27.1	44.5	70.4

By drawing a suitable linear graph, estimate the values of a and b, giving both answers to one decimal place. *(9 marks)*

(JMB)

9 Referred to an origin O and coordinate axes Ox and Oy, a curve is given by

$$x = \sec t + \tan t, \quad y = \operatorname{cosec} t + \cot t, \quad 0 < t < \pi/2,$$

where t is a parameter.

Prove that $\dfrac{dy}{dx} = \dfrac{1 - \sin t}{1 - \cos t}$. *(5 marks)*

Show that the normal to the curve at the point S, where $t = \tan^{-1}\left(\frac{3}{4}\right)$, has equation $x - 2y + 4 = 0$. Find an equation of the normal to the curve at the point T, where $t = \tan^{-1}\left(\frac{4}{3}\right)$. These normals meet at the point N. Find the coordinates of N. *(5 marks)*

Hence calculate

a) the area of the triangle SNT, *(3 marks)*
b) the tangent of the angle SNT. *(3 marks)*

(AEB, 1986)

10 a) Integrate the following with respect to x:

i) $\dfrac{1}{1+x}$ ii) $\dfrac{2+x}{1+x}$ iii) xe^x.

b) If $x = 2\cos^2\theta + 5\sin^2\theta$, show that $x - 2 = 3\sin^2\theta$. Hence, or otherwise, evaluate

$$\int_2^5 \frac{dx}{\sqrt{(x-2)(5-x)}}.$$ (SU)

Paper 8

1 a) Find the set of values of k for which

$$x^2 + (k-2)x + 2k + 1$$

is positive for all real values of x.   *(4 marks)*

b) The roots of the equation $x^2 + (k-2)x + 2k + 1 = 0$ are α and β. Find a quadratic equation with coefficients in terms of k whose roots are $\alpha^2 + \alpha$ and $\beta^2 + \beta$.   *(7 marks)*

Given that $\alpha^2 + \alpha = \beta^2 + \beta$ and that $\alpha \neq \beta$, find the value of k and the double root of your equation.   *(5 marks)*
    (AEB, 1987)

2 The quadratic expression $f(x)$ is such that
a) when $f(x)$ is divided by $x - 1$, the remainder is 3;
b) when $f(x)$ is divided by $x + 1$, the remainder is 7;
c) $f(0) = 1$.
Find $f(x)$.   (O&C/MEI)

3 Given that $\frac{31}{20}$ is a rational approximation to the real root of the equation

$$\log_e x + x - 2 = 0$$

use the Newton–Raphson method to locate the root, correct to three decimal places.
    (NI)

4 In a model to estimate the depreciation of the value of a car, it is assumed that the value, £V, at age t months, decreases at a rate which is proportional to V. Using this model, write down a differential equation relating V and t. Given that the car has an initial value of £6000, solve the differential equation and show that

$$V = 6000e^{-kt},$$

where k is a positive constant.

The value of the car is expected to decrease to £3000 after 36 months. Calculate

a) the value, to the nearest pound, of the car when it is 15 months old,

b) the age of the car, to the nearest month, when its value is £2000.

(JMB)

5 A chord AB of a circle divides the area of that circle in the ratio 1 : 2.

If O is the centre of the circle and angle AOB = θ radians, prove that $\sin \theta = \theta - \dfrac{2\pi}{3}$.

By drawing, accurately, the graphs of $y = \sin \theta$ and $y = \theta - \dfrac{2\pi}{3}$, between $\theta = 0$ and $\theta = \pi$ on the same diagram, find the angle AOB in radians.

(SU)

6 A pyramid has a square base ABCD of side a and its vertex is E. Each sloping face makes an angle $\dfrac{\pi}{3}$ with the base.

a) Find the height of the pyramid.

b) Find the length of each sloping edge.

c) Find the angle between each sloping edge and the base.

d) Show that

$$\sin E\widehat{A}B = \frac{2}{\sqrt{5}}.$$

(7 marks)

P is a point on BE such that AP and CP are both perpendicular to BE.

e) Calculate the length of AP.

f) Show that the angle between adjacent sloping faces is $\cos^{-1}\left(-\tfrac{1}{4}\right)$.

(6 marks)

Is it possible to construct a pyramid on a square base whose adjacent sloping faces are perpendicular? Explain your answer briefly. (2 marks)

(W)

7 A curve has equation $y = f(x)$ where $f(x) = \dfrac{x^2 + 5}{x - 2}$. Find $\dfrac{dy}{dx}$. Determine the nature and coordinates of any stationary points.

Sketch the graphs of a) $y = f(x)$
   b) $y = f(|x|)$.

(16 marks)
(O&C/MEI)

8 a) Differentiate $\dfrac{1}{x}$ from *first principles* (with respect to x).

   b) Show that, when $x = \sqrt{2}$, the derivative (with respect to x) of

$$\frac{1}{x} \sin^{-1}\left(\frac{1}{x}\right)$$

has the value $-\frac{1}{8}(4 + \pi)$.

(NI)

9 a)

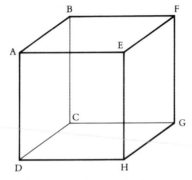

The above figure shows a cube. Calculate, to the nearest degree, the angles between
   i)   AG and EF, *(2 marks)*
   ii) AG and BH, *(2 marks)*
   iii) AG and the plane ABFE. *(2 marks)*

   b) The points A, B, C have position vectors

$$\begin{aligned}
\mathbf{OA} &= -2\mathbf{i} + 3\mathbf{j} - 7\mathbf{k} \\
\mathbf{OB} &= \mathbf{i} + 7\mathbf{j} + 5\mathbf{k} \\
\mathbf{OC} &= 4\mathbf{i} + 3\mathbf{j} + \mathbf{k}
\end{aligned}$$

relative to an origin O and a set of mutually perpendicular unit vectors $\mathbf{i}, \mathbf{j}, \mathbf{k}$. Show that

$$\mathbf{AB} = 3\mathbf{i} + 4\mathbf{j} + 12\mathbf{k}$$

and obtain a similar expression for $\mathbf{AC}$. Calculate $|\mathbf{AB}|$, $|\mathbf{AC}|$ and the scalar product $\mathbf{AB}.\mathbf{AC}$ and deduce that

$$B\widehat{A}C = \cos^{-1}\left(\tfrac{57}{65}\right).$$

Write down an expression for the position vector of the point D which divides AC in the ratio $\lambda : 1 - \lambda$. By equating the scalar product $\mathbf{AC}.\mathbf{OD}$ to zero, deduce that the position vector of the foot of the perpendicular from O to AC is given by

$$\tfrac{52}{25}\mathbf{i} + 3\mathbf{j} - \tfrac{39}{25}\mathbf{k}.$$

(9 marks)
(W)

10 Show that the equation $2x^3 + 3x^2 - 4 = 0$ has only one real root and that this root lies between $x = 0$ and $x = 1$. Using the Newton–Raphson method and a starting value of 1, find the value of the root correct to 3 decimal places. Explain why you are confident that you have achieved the required accuracy.

An alternative iterative process

$$x_{n+1} = \frac{1}{2}\left[x_n + \sqrt{\left(\frac{4 - 2x_n^3}{3}\right)}\right], \quad n \geqslant 0,$$

is proposed. Show algebraically that, if this converges, it does so to the real root of $2x^3 + 3x^2 - 4 = 0$.

Investigate whether this process converges with starting values of
a) 1,
b) 2. (O&C/MEI)

Paper 9

1 Prove by induction that the following results are true for all positive integers n.

a) $\displaystyle\sum_{r=1}^{n} r^2(r - 1) = \frac{1}{12}n(n^2 - 1)(3n + 2)$.

b) Given that $y = xe^x$, then $\dfrac{d^n y}{dx^n} = (x + n)e^x$. (C)

2 Determine all of the angles between $0°$ and $180°$ which satisfy the equation

$$\cos 5\theta + \cos 3\theta = \sin 3\theta + \sin \theta.$$

(NI)

3 Sketch, in the same diagram, the graphs for $x > 0$ of $y = \ln x$ and $y = \dfrac{2}{x}$.

Use the Newton–Raphson process to find, correct to two decimal places, the x-coordinate of the point where the two graphs meet.

(O&C)

4 Prove that the tangent at P $(4, 4)$ to the curve $4y = x^2$ has equation

$$2x - y - 4 = 0. \qquad \text{(3 marks)}$$

This tangent meets the line $4x + 3y - 12 = 0$ at the point Q.
Calculate the coordinates of Q. *(2 marks)*

The normal at P to the curve $4y = x^2$ meets the line $4x + 3y - 12 = 0$ at the point R.
Calculate the coordinates of R. *(4 marks)*

Show that the circle on QR as diameter has equation

$$x^2 + (y - 4)^2 = 16. \qquad \text{(3 marks)}$$

Sketch in the same diagram the curve $4y = x^2$ and the circle $x^2 + (y - 4)^2 = 16$ and write on your sketch the coordinates of any points where the two curves meet or intersect.
(4 marks)
(AEB, 1987)

5 Sketch the graph of $y = \sec\left(x - \dfrac{\pi}{4}\right)$ for $0 \leqslant x \leqslant \dfrac{\pi}{2}$, stating the y-coordinates corresponding to $x = 0$, $x = \dfrac{\pi}{4}$, $x = \dfrac{\pi}{2}$.

The region bounded by this graph, the x-axis and the lines $x = 0$, $x = \dfrac{\pi}{2}$, is rotated through one revolution about the x-axis. Find the volume generated.
(JMB)

6 Verify that, for any value of A and $x \neq 0$,

$$y = Ax^2 + \frac{1}{x^2}$$

satisfies the differential equation

$$\frac{dy}{dx} = \frac{2y}{x} - \frac{4}{x^3}.$$

Hence find the solution (for $x > 0$) of the differential equation for which $y = 1$ when $x = 2$.

(O&C/SMP)

7 a) In an arithmetic progression, the sum of the first four terms is 22 and the sum of the first five terms is 35. Find the hundredth term. *(4 marks)*
  b) In a geometric progression, the first term is 12 and the fourth term is $-\frac{3}{2}$. Find the sum, S_n, of the first n terms of the progression. *(5 marks)*

  Find the sum to infinity, S, of the progression and the least value of n for which the magnitude of the difference between S_n and S is less than 0.001. *(3 marks)*
(C)

8 a) The length a of the side BC of a triangle ABC is related to the lengths b, c of the sides CA, AB and the included angle $\widehat{BAC}$ (measured in radians) by the cosine formula

$$a^2 = b^2 + c^2 - 2bc \cos A.$$

If A is increased by 1% and b, c held constant, show that a increases by x% where

$$x \approx \frac{Abc \sin A}{a^2}.$$

(6 marks)

b)

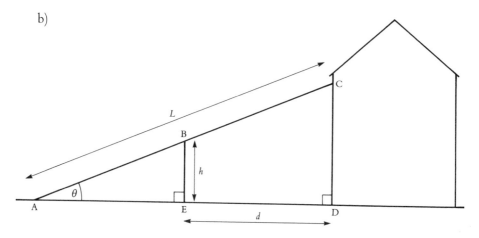

The above figure shows a ladder AC of length L leaning against the side of a house. The ladder, which is inclined at an angle θ to the horizontal AD, also rests on the top of a wall BE of height h situated at a distance d from the house. Obtain an expression for L in terms of h, d and θ and show that its minimum value as θ varies is

$$(h^{\frac{2}{3}} + d^{\frac{2}{3}})^{\frac{3}{2}}$$

(9 marks)
(W)

9 The functions f and g are defined by

$$f: x \mapsto x^2 - 2, \text{ where } x \in \mathbb{R},$$
$$g: x \mapsto e^x, \text{ where } x \in \mathbb{R}.$$

Write down the composite function $f \circ g$ and the inverse function g^{-1}, stating the domain of each.

Sketch, in separate diagrams, the graphs of $f \circ g$ and g^{-1}, showing the general shape of each graph and writing on your sketches the coordinates of points where the curves cross the coordinate axes.

(7 marks)
(AEB, 1986)

125

10 a) Use integration by parts to evaluate

$$I = \int_0^{\frac{\pi}{6}} x \sin x \, dx.$$

b) Deduce that

$$8 \int_0^{\frac{\pi}{12}} x \sin x \cos x \, dx = I. \qquad \text{(NI)}$$

Paper 10

1 Use identities for $\cos (C + D)$ and $\cos (C - D)$ to prove that

$$\cos A + \cos B = 2 \cos \frac{A + B}{2} \cos \frac{A - B}{2}. \qquad \textit{(2 marks)}$$

Hence find, in terms of π, the general solution of the equation

$$\cos 5\theta + \cos \theta = \cos 3\theta. \qquad \textit{(5 marks)}$$

Using both the identity for $\cos A + \cos B$, and the corresponding identity for $\sin A - \sin B$, show that

$$\sin 5\alpha - \sin \alpha = 2 \sin \alpha \, (\cos 4\alpha + \cos 2\alpha). \qquad \textit{(3 marks)}$$

The triangle PQR has angle QPR $= \alpha$ (which is not zero), angle PQR $= 5\alpha$ and RP $= 3$RQ. Show that $\sin 5\alpha = 3 \sin \alpha$ and deduce that

$$\cos 4\alpha + \cos 2\alpha = 1. \qquad \textit{(3 marks)}$$

By solving a quadratic equation in $\cos 2\alpha$, or otherwise, find the value of α, giving your answer to the nearest one tenth of a degree. *(3 marks)*
(AEB, 1986)

2 By writing $e^x = u$, or otherwise, solve the equation

$$e^x - 2e^{-x} = 1. \qquad \text{(C)}$$

3 Real functions f, g, k are described by

$$f(x) = x^2 - 1, \quad g(x) = 2x + 4, \quad k(x) = \sqrt{2x + 5}.$$

a) Show that, for all $x \geqslant -\frac{5}{2}$, $f(k(x)) = g(x)$.

b) Describe the real function h which satisfies

$$g \circ h = f$$

where $(g \circ h)(x) = g(h(x))$ for all real x.

c) Show that k $\circ$ h is the modulus function. (NI)

4 a) The variables p and q are related by the law

$$q = ap^b, \text{ where } a \text{ and } b \text{ are constants.}$$

Given that $\ln p = 1.32$ when $\ln q = 1.73$,
and $\ln p = 0.44$ when $\ln q = 1.95$ find the values of b and $\ln a$.

(5 marks)

b) Given that $y = \log_2 x$ and that

$$\log_2 x - \log_x 8 + \log_2 2^k + k \log_x 4 = 0,$$

prove that

$$y^2 + ky + (2k - 3) = 0. \qquad \text{(4 marks)}$$

i) Hence deduce the set of values of k for which y is real. *(4 marks)*

ii) Find the values of x when $k = 1.5$. *(3 marks)*

(AEB, 1986)

5 A circle S is given by the equation

$$x^2 + y^2 - 4x + 6y - 12 = 0.$$

Find the radius of S and the coordinates of the centre of S. Calculate the length of the perpendicular from the centre of S to the line L whose equation is

$$3x + 4y = k,$$

where k is a constant. Deduce the values of k for which L is a tangent to S.

(JMB)

6 Two points P and Q on the parabola $y^2 = 4ax$ have coordinates $(ap^2, 2ap)$ and $(aq^2, 2aq)$ respectively. Find

a) the slope of the chord PQ,

b) the slope of the tangent at P to the parabola,

c) the equation of the normal at P to the parabola.

The normals at P and Q to the parabola intersect at R. Find the coordinates of R.

If the chord passes through the point $(-2a, 0)$ show that R lies on the parabola.

(O&C)

7 Using the substitution $x = \dfrac{1}{t}$, find $\displaystyle\int_2^3 \dfrac{dx}{x(x^2 - 1)^{\frac{1}{2}}}$ correct to 2 decimal places.

(O&C/MEI)

8 The position vectors of the points A, B and C, relative to a fixed origin O, are $\mathbf{a} = 6\mathbf{i} + 4\mathbf{j} - \mathbf{k}$, $\mathbf{b} = 8\mathbf{i} + 5\mathbf{j} - 3\mathbf{k}$ and $\mathbf{c} = 2\mathbf{i} + 8\mathbf{j} - 5\mathbf{k}$, respectively.

Find a) the vector $\overrightarrow{AB}$, *(2 marks)*

    b) the length of $\overrightarrow{AB}$, *(2 marks)*

    c) the cosine of angle ABC, *(3 marks)*

    d) the area of triangle ABC. *(3 marks)*

Show that, for all values of the parameter t, the point P with position vector $\mathbf{p} = (8 + 2t)\mathbf{i} + (5 + t)\mathbf{j} - (3 + 2t)\mathbf{k}$ lies on the straight line passing through A and B.

Determine the value of t for which OP is perpendicular to AB and hence, or otherwise, calculate the shortest distance from O to the line AB. *(6 marks)*
(AEB, 1987)

9 On three separate Argand diagrams, sketch clearly the regions of the complex plane for which

a) $|z| \leqslant 2$,   b) $\text{Re}\,(z) \geqslant 1$,   c) $-\dfrac{\pi}{3} \leqslant \arg z \leqslant \dfrac{\pi}{3}$.

On a fourth diagram, sketch the region D where all the inequalities are satisfied and state which inequality is not necessary in defining the region D.

Find in the form $a + bj$, the complex numbers representing points in the region D for which $|z - z^*|$ has a maximum value, and state this value.

$(z^*$ is the conjugate of z.) *(16 marks)*
(O&C/MEI)

10

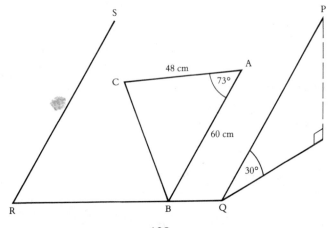

The diagram shows a plane PQRS inclined at 30° to the horizontal. The line AB is 60 cm long and runs down a line of greatest slope. The point C is on the inclined plane; the length of AC is 48 cm and the angle CAB is 73°. Find

a) the length of BC, to the nearest centimetre,
b) the angle ABC, to the nearest degree.

The point D is the foot of the perpendicular from C to AB. Show that the vertical height of D above B is 23 cm to the nearest centimetre. Find, to the nearest degree, the angle between BC and the horizontal plane.

(JMB)

Paper 11

1 Show that

$$\sin 2x + \sin 4x + \sin 6x \equiv \sin 4x \,(1 + 2\cos 2x).$$

Hence prove the identity

$$\sin 3x \sin 4x \equiv (\sin 2x + \sin 4x + \sin 6x)\sin x.$$

Deduce that $\sin\left(\dfrac{\pi}{12}\right) = \dfrac{1}{\sqrt{6 + \sqrt{2}}}$.

(7 marks)
(AEB, 1986)

2 The matrix $M = \begin{pmatrix} a & b \\ c & d \end{pmatrix}$ maps the point $\begin{pmatrix} 2 \\ 1 \end{pmatrix}$ on to the point $\begin{pmatrix} 1 \\ 2 \end{pmatrix}$ in the x, y plane. Find an equation relating a and b and an equation relating c and d. If in addition $\det M = 1$, show that $d = 2 + 2b$ and find a and c in terms of b.

Find the value of b such that M represents a rotation about the origin. Show that the angle of rotation is approximately 37°.

(16 marks)
(O&C/MEI)

3 The base of a pyramid is a square of side a. Each face makes an angle of $\dfrac{\pi}{3}$ radians with the base. Show that

a) the height of the pyramid is $\dfrac{a\sqrt{3}}{2}$, and

b) the angle between two adjacent sloping faces is $\cos^{-1}\left(-\tfrac{1}{4}\right)$ radians.   (NI)

4 The lines L and M are given respectively by the equations

$$\mathbf{r} = \begin{pmatrix} 2 \\ -3 \\ 1 \end{pmatrix} + s \begin{pmatrix} 1 \\ 2 \\ 2 \end{pmatrix} \quad \text{and} \quad \mathbf{r} = \begin{pmatrix} 8 \\ 5 \\ 13 \end{pmatrix} + t \begin{pmatrix} 3 \\ 2 \\ 6 \end{pmatrix}.$$

Show that L and M intersect and find the position vector of A, their point of intersection.

Verify that L and M both lie in the plane Π given by the equation

$$\mathbf{r} \cdot \begin{pmatrix} 2 \\ 0 \\ -1 \end{pmatrix} = 3.$$

The point B is $(12, 5, 6)$ and the point C is the foot of the perpendicular from B to Π. Find a vector equation for BC and hence find the position vector of C. Show that C lies on L. *(13 marks)*

(JMB)

5 The complex number z satisfies the equation

$$2zz^* - 4z = 3 - 6i,$$

where z^* is the complex conjugate of z. Find, in the form $x + iy$, the two possible values of z.

(JMB)

6 a)

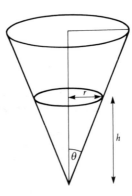

The figure shows a hollow inverted right circular cone with semi-vertical angle θ. Water is poured into the cone and at time t, the solid cone of water has volume V, height h, base radius r and base area A. Given that the volume of water increases at a constant rate k per unit time, show that

$$\frac{dh}{dt} = \frac{k}{\pi r^2} \quad \text{and} \quad \frac{dA}{dt} = \frac{2k}{h}. \qquad \text{(7 marks)}$$

b) A sector of a circle has radius r and the angle between the two radii is θ. Obtain expressions for the area A and perimeter P of the sector in terms of r and θ and show that

$$A = \tfrac{1}{2}r(P - 2r).$$

Hence show that the maximum area of a sector with a given perimeter P is $\frac{1}{16}P^2$.

Find the minimum perimeter of a sector with a given area A.   (8 marks)
(W)

7 a) Show that the sum of the first 50 terms of the geometric series with first term 1, and common ratio $r = 2^{\frac{1}{50}}$ is $\dfrac{1}{r-1}$.

Hence, by using the trapezium rule with 50 intervals of equal width, show that $\displaystyle\int_0^1 2^x\,dx$ is approximately $\dfrac{r+1}{100(r-1)}$, where $r = 2^{\frac{1}{50}}$. Evaluate this expression to 3 decimal places.   (12 marks)

b) By writing 2^x as $e^{x\ln 2}$, find the exact value of $\displaystyle\int_0^1 2^x\,dx$, leaving your answer in terms of $\ln 2$.
(4 marks)
(AEB, 1986)

8 At time t, the position of a particle is given by

$$\mathbf{r} = 2(1 + \cos t)\mathbf{i} + 2(t + \sin t)\mathbf{j}$$

where $\mathbf{i}$ and $\mathbf{j}$ are mutually perpendicular unit vectors. Show that, at time t, the velocity of the particle has magnitude $4\left|\cos\left(\frac{t}{2}\right)\right|$.

What is the magnitude of its acceleration at any time t? At what times t is the acceleration perpendicular to the velocity?

(NI)

9 Sketch the graphs of $y = |x - 1|$ and $y = x^2 - 1$ using the same axes. Hence, or otherwise, find all the values of x for which $|x - 1| < x^2 - 1$.
(7 marks)
(O&C/MEI)

10 a) Evaluate

$$\int_2^4 \frac{1}{2x^2 + x}\, dx,$$

$$\int_0^{\frac{\pi}{4}} \sin 3x \cos 2x\, dx,$$

giving your answer in each case to three decimal places.

b) Find $\int (x^2 + 1)e^{2x}\, dx.$

c) Using the substitution $x = 2(\sec\theta - 1)$, or otherwise, find

$$\int \frac{1}{(x + 2)(x^2 + 4x)^{\frac{1}{2}}}\, dx. \qquad \text{(O\&C)}$$

Paper 12

1 In the expansion of

$$\frac{1}{\sqrt{(1 + ax)}} - \frac{1}{1 + 2x}$$

in ascending powers of x, the first non-zero term is the term in x^2. Find the value of the constant a and hence find the terms in x^2 and x^3. *(7 marks)*
(JMB)

2 The figure shows three circular discs with centres A, B and C and radii 3 cm, 2 cm and 1 cm respectively touching each other externally.

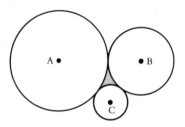

Show that the tangent of angle BAC is $\frac{3}{4}$. *(2 marks)*

Find the area of the shaded region enclosed by the three discs, giving your answer in cm^2 to 3 significant figures. *(6 marks)*
(AEB, 1987)

3 The circle with equation $(x-5)^2 + (y-7)^2 = 25$ has centre C. The point P (2, 3) lies on the circle. Determine the gradient of PC and hence, or otherwise, obtain the equation of the tangent to the circle at P. *(4 marks)*

Find also the equation of the straight line which passes through the point C and the point Q (−1, 4). The tangent and the line CQ intersect at R. Determine the size of angle PRC, to the nearest 0.1°. *(4 marks)*

(AEB, 1987)

4 Show that the equation $x^3 + 2x - 1 = 0$ has only one real root α, and that $0 < \alpha < 1$. Using the iterative procedure $x_{n+1} = \frac{1}{2}(1 - x_n^3)$ with a starting value $x_0 = 0.5$, find the value of α correct to 2 decimal places. *(7 marks)*

(O&C/MEI)

5 Prove the identity

$$\sqrt{3} \cos \theta + \sin \theta \equiv 2 \cos \left(\theta - \tfrac{\pi}{6}\right).$$ *(3 marks)*

a) Find, in terms of π, the general solution of the equation

$$\sqrt{3} \cos \theta + \sin \theta = 1.$$ *(4 marks)*

b) A curve has equation $y = \dfrac{1}{\sqrt{3} \cos x + \sin x}$.

The region enclosed by the curve, the coordinate axes, and the line $x = \tfrac{\pi}{6}$ is denoted by R.

i) Prove that the area of R is $\tfrac{1}{4} \ln 3$. *(5 marks)*

ii) Find, in terms of π, the volume generated by rotating R completely about the x-axis. *(4 marks)*

(AEB, 1986)

6 For the curve whose equation is

$$y = \frac{x}{x^2 + 1}$$

show that

a) $y = x$ is the equation of the tangent at the origin;

b) $(1, \tfrac{1}{2})$ is a maximum turning point;

c) $\left(\sqrt{3}, \dfrac{\sqrt{3}}{4}\right)$ is a point of inflexion:

d) the x-axis is an asymptote;

e) the curve is symmetrical about the origin.

Sketch the curve.

(NI)

7 Sketch the curve C given by the equation

$$xy = 1.$$

A line of gradient m, where $m \neq 0$, passes through the point A $(3, 0)$ and meets C at the points P and Q. Show that the x-coordinates of P and Q are the roots of the equation

$$mx^2 - 3mx - 1 = 0.$$

a) Find the set of values of m for which this equation has real unequal roots. Find also the (non-zero) value of m for which the roots are equal, and hence, or otherwise, find the equation of the tangent to C from the point A.

b) Show that the mid-point M of PQ lies on the line $x = \dfrac{3}{2}$. Show on your diagram of C the line $x = \dfrac{3}{2}$ and the possible positions of M on this line. Indicate that part of the line which corresponds to positive values of m.

(JMB)

8 a) Obtain the coordinates of the turning point on the curve with equation $y = x^2 \ln x$, where $x > 0$, and find whether it is a maximum point or a minimum point.

(6 marks)

b)

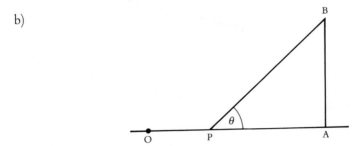

In the diagram, O and A are fixed points 1000 m apart on horizontal ground. The point B is vertically above A, and represents a balloon which is ascending at a steady rate of 2 m s^{-1}. The balloon is being observed from a moving point P on the line OA. At time $t = 0$, the balloon is at A and the observer is at O. The observation point P moves towards A with steady speed 6 m s^{-1}. At time t, the angle APB is θ radians. Show that

$$\frac{d\theta}{dt} = \frac{500}{t^2 + (500 - 3t)^2}.$$

(6 marks)

(C)

134

9 Show that the curve

$$y = \sqrt{\frac{x}{3-x}}$$

increases steadily from $x = 0$ to $x = \frac{3}{2}$. *(5 marks)*

By using the substitution $x = 3\sin^2 t$, or otherwise, show that the area of the region bounded by the curve, the x-axis, and the ordinate $x = \frac{3}{2}$, is

$$\tfrac{3}{4}(\pi - 2). \qquad \text{(10 marks)}$$

Evaluate the volume of the solid generated when the above region is rotated about the x-axis through an angle 2π radians. *(9 marks)*

(NI)

10 Showing your working in the form of a table, use Simpson's rule with 4 equal intervals to estimate

$$\int_0^2 \ln\left(1 + x^2\right) \, dx,$$

giving your answer to 3 decimal places. *(4 marks)*

Deduce an approximate value for

$$\int_0^2 \ln\sqrt{\left(1 + x^2\right)} \, dx. \qquad \text{(2 marks)}$$

(AEB, 1987)

Paper 13

1 Given $x = 8 - 27y^3$, find y in terms of x. Hence find the first three terms in the expansion of y in terms of x for small x. *(O&C/MEI)*

2 Use induction to prove that the sum to n terms of the series

$$(1 \times 3 \times 5) + (2 \times 4 \times 6) + (3 \times 5 \times 7) + \ldots$$

is

$$\tfrac{1}{4}n(n+1)(n+4)(n+5). \qquad \text{(O&C)}$$

3 Show that, if x is real, then $x^2 + 6x + 14$ is always positive.

Hence, or otherwise, determine the real values of x for which

a) $\dfrac{(x+3)(x+4)}{(x+1)} > \dfrac{x+8}{3};$

b) $\dfrac{(x+3)(x+4)}{(x+1)(x+8)} > \dfrac{1}{3}.$

(NI)

4 The functions f and g are defined by

$$f: x \mapsto x^2 - 3, \quad x \in \mathbb{R},$$

$$g: x \mapsto 2x + 5, \quad x \in \mathbb{R}.$$

Find in a similar form the composite function f ∘ g.

Sketch on separate axes the graphs of f and f ∘ g.

Hence, or otherwise, show that the range of f corresponding to the domain $-4 \leqslant x \leqslant 4$ is $-3 \leqslant f(x) \leqslant 13$, and find the range of f ∘ g corresponding to this domain.

(8 marks)

(AEB, 1986)

5 The complex number z has modulus 4 and argument $\dfrac{\pi}{6}$.

The complex number w has modulus 2 and argument $-\dfrac{2\pi}{3}$.

a) Express z and w in the form $a + ib$, where a and b are integers or surds.
b) Mark in an Argand diagram the points P and Q which represent z and w, respectively. Calculate PQ^2, giving your answer in the form $r + s\sqrt{3}$, where r and s are integers.
c) Find the modulus and the argument of

$$\frac{z}{w} \quad \text{and} \quad w^2,$$

giving each argument between $-\pi$ and π.

(9 marks)

(JMB)

6 a) Show that

$$\tan 4\theta = \frac{4 \tan \theta \, (1 - \tan^2 \theta)}{\tan^4 \theta - 6 \tan^2 \theta + 1}.$$

Hence show that $\tan 22\frac{1}{2}°$ is a root of the equation

$$t^4 - 6t^2 + 1 = 0.$$

Deduce that

$$\tan 22\frac{1}{2}° = \sqrt{3 - \sqrt{8}}$$

and obtain a similar expression for $\tan 67\frac{1}{2}°$.

Use your results to show that

$$\sec^2 22\frac{1}{2}° + \sec^2 67\frac{1}{2}° = 8.$$

(8 marks)

b) AB is a chord of a circle centre O which subtends an angle 2θ radians at O $(\theta < \frac{1}{2}\pi)$. If AB divides the circle into two regions, one having twice the area of the other, show that θ satisfies the equation

$$6\theta - 3 \sin 2\theta - 2\pi = 0.$$

Hence show that θ lies between 1.30 and 1.31.

(7 marks)

(W)

7 Sketch the graph of $y = (x - 2)^2 - 4$, showing clearly on your graph the coordinates of any stationary points and of the intersections with the axes.

(3 marks)

Find the coordinates of the stationary points on the graph of

$$y = (x - 2)^3 - 12(x - 2)$$

and sketch the graph, giving the exact coordinates (in surd form, where appropriate) of the intersections with the axes. *(5 marks)*

Find the set of values of x for which

$$(x - 2)^3 > 12(x - 2).$$ *(2 marks)*

Find the set of values of k for which the equation

$$(x - 2)^3 - 12(x - 2) + k = 0$$

has exactly one real root.

(2 marks)

(C)

8 i, j and **k** are unit vectors parallel to the x, y and z axes of a Cartesian frame of reference Oxyz, origin O.

A line L$_1$, passes through the point $(3, 6, 1)$ and is parallel to the vector $2\mathbf{i} + 3\mathbf{j} - \mathbf{k}$.

A line L$_2$, passes through the point $(3, -1, 4)$ and is parallel to the vector $\mathbf{i} - 2\mathbf{j} + \mathbf{k}$.

a) Using the form $\mathbf{r} = \mathbf{a} + \mathbf{b}t$, write down the vector equations of the lines L$_1$ and L$_2$.

Show that the lines intersect and find the coordinates of the point of intersection. *(8 marks)*

b) What is the acute angle between the lines? *(4 marks)*

c) The point A $(5, 9, 0)$ lies on L$_1$ whilst the point B $(5, a, b)$ lies on L$_2$. Find a and b and hence find the point C which lies on the line AB such that AC : CB $= 1 : 2$.

What is the magnitude of $\overrightarrow{OC}$?

What is the unit vector parallel to $\overrightarrow{OC}$? *(12 marks)*

(NI)

9 Given that $\dfrac{7x - x^3}{(2 - x)(x^2 + 1)} \equiv \dfrac{A}{(2 - x)} + \dfrac{Bx + C}{(x^2 + 1)}$, determine the values of A, B, C.

(4 marks)

A curve has equation $y = \dfrac{7x - x^3}{(2 - x)(x^2 + 1)}.$

Determine the equation of the normal to the curve at the point $(1, 3)$.

(5 marks)

Prove that the area of the region bounded by the curve, the x-axis and the line $x = 1$ is $\dfrac{7}{2} \ln 2 - \dfrac{\pi}{4}$.

(7 marks)

(AEB, 1987)

137

10 The integral

$$\int_0^1 \frac{x}{\sqrt{(4-x^2)}}\, dx$$

can be found by using either of the substitutions

a) $x = 2 \sin \theta$,   b) $u = 4 - x^2$.

Find the value of the integral by both methods, and check that they give the same answer.   (O&C/SMP)

Paper 14

1 Express $(12 \cos \theta - 5 \sin \theta)$ in the form $R \cos (\theta + \alpha)$, where $R > 0$. The function $g(x)$ is defined by

$$g(x) = 27 \cos^2 x - 10 \sin x \cos x + 3 \sin^2 x.$$

Express $g(x)$ in the form $a \cos 2x + b \sin 2x + c$, where a, b, c are constants to be determined.

Find

a) the greatest and least values of $g(x)$,

b) the values of x in the range $0° < x < 360°$ which satisfy the equation $g(x) = 2$,

c) the range of values of x within the range $0° < x < 180°$ for which

$$4 \leqslant 2g(x) \leqslant 17.$$

[All angles should be given to the nearest $0.1°$.]   (O&C)

2 A circular sector of area A cm^2, has bounding radii, each of length x cm, and the angle between these radii is θ radians. Given that the perimeter of the sector is 12 cm,

a) express θ in terms of x,   *(2 marks)*

b) show that $A = 6x - x^2$,   *(2 marks)*

c) find the greatest value of A as x varies.   *(2 marks)*
                                                                           (AEB, 1986)

3

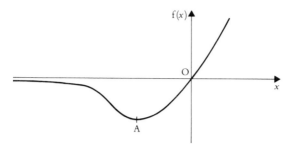

The figure shows the shape of the graph of the function f, where

$$f: x \to xe^x, \quad x \in \mathbb{R}.$$

a) Determine the coordinates of the stationary point A and hence write down the range of f. *(4 marks)*

b) Sketch the graph of curve $y = (x - 2)e^{x-2}$ and mark on the sketch the coordinates of any points where the curve crosses the coordinate axes.

(3 marks)

(AEB, 1987)

4 A Cartesian frame of reference, having origin O, is defined by the mutually perpendicular unit vectors **i**, **j** and **k**.

a) Find the vector equation of the line which passes through the two points, having position vectors relative to O, $2\mathbf{i} - \frac{1}{3}\mathbf{j} + 2\mathbf{k}$ and $\mathbf{i} + \frac{1}{3}\mathbf{j}$ in the form $\mathbf{r} = \mathbf{a} + t\mathbf{b}$.

b) A second line has vector equation

$$\mathbf{r} = -3\mathbf{i} - 4\mathbf{j} - \tfrac{1}{2}\mathbf{k} + s(2\mathbf{i} + \mathbf{j} + \tfrac{3}{2}\mathbf{k}).$$

Find the point of intersection of the two lines. *(NI)*

5 Derive the equation of the normal to the parabola $y^2 = 4ax$ at the point $(at^2, 2at)$. *(4 marks)*

Three normals to the parabola $y^2 = 8x$ each pass through the point $(12, 0)$. Determine their equations. *(8 marks)*

(NI)

6 The parametric coordinates of a point on the ellipse $\dfrac{x^2}{a^2} + \dfrac{y^2}{b^2} = 1$ are $x = a \cos t, y = b \sin t$.

Find $\dfrac{dy}{dx}$ in terms of a) t,

b) x and y.

Hence, or otherwise, show that the equation of the tangent at (x_1, y_1) can be written $\dfrac{xx_1}{a^2} + \dfrac{yy_1}{b^2} = 1$.

Prove that $y^3 \left(\dfrac{d^2y}{dx^2} \right) = -\dfrac{b^4}{a^2}$. *(16 marks)*

(O&C/MEI)

7 A rainwater-butt has a height of 100 cm and a uniform cross-sectional area of 2000 cm². At a time when the butt is full of water it begins to leak from a small hole in the base. The depth of the water which remains t minutes after the leak begins is x cm. Given that the water leaks out at the rate of $100\sqrt{x}$ cm³ per minute, and that no water enters the butt, show that

$$\frac{dx}{dt} = -\frac{1}{20}\sqrt{x}.$$

When the leak is first noticed, the butt is found to be only half full. Find, to the nearest minute, the time for which the butt has been leaking. *(8 marks)*
(JMB)

8 Given that

$$y = \frac{3x - 14}{(x - 2)(x + 6)},$$

express y as a sum of partial fractions. Hence find $\dfrac{dy}{dx}$ and $\dfrac{d^2y}{dx^2}$. Show that $\dfrac{dy}{dx} = 0$ for $x = 10$ and for one other value of x.

Find the maximum and minimum values of y, distinguishing between them.
(JMB)

9 Sketch, on the same diagram, for $x \geqslant 0$, the graphs of $y = x^{\frac{1}{2}}$ and $y = x^{\frac{3}{2}}$, labelling each graph clearly.

[An accurate plot is not required. You should show the general shape of each curve particularly near the origin, and give the coordinates of the point(s) of intersection of the curves.] *(4 marks)*

Show that $x = \frac{1}{2}$ satisfies the equation

$$x^{\frac{1}{2}} = 2x^{\frac{3}{2}}$$

and sketch also on your diagram the curve $y = 2x^{\frac{3}{2}}$. *(3 marks)*

Show, by integration, that the area of the finite region between the curves $y = x^{\frac{1}{2}}$ and $y = 2x^{\frac{3}{2}}$ is $\dfrac{2^{\frac{1}{2}}}{15}$. *(5 marks)*
(C)

10 Given that $x < 4$, find

$$\int \frac{8}{(4-x)(8-x)} \, dx.$$

A chemical reaction takes place in a solution containing a substance S. At noon there are two grams of S in the solution and t hours later there are x grams of S. The rate of the reaction is such that x satisfies the differential equation

$$8 \frac{dx}{dt} = (4-x)(8-x).$$

Solve this equation, giving t in terms of x.

Find, to the nearest minute, the time at which there are three grams of S present. *(10 marks)*

(JMB)

Paper 15

1 Given that p and q are real and that $1 + 2i$ is a root of the equation

$$z^2 + (p + 5i)z + q(2 - i) = 0$$

determine a) the values of p and q, *(5 marks)*

b) the other root of the equation. *(2 marks)*

(AEB, 1987)

2 Prove by induction, or otherwise, that

$$1 \times 5 + 3 \times 7 + 5 \times 9 + \ldots + (2n - 1)(2n + 3) = \frac{n}{3}(4n^2 + 12n - 1),$$

for any positive integer n. *(6 marks)*

(O&C/MEI)

3 Solve each of the following equations, to find x in terms of a, where $a > 0$ and $a \neq e^2$:

a) $a^x = e^{2x + 1}$,

b) $2 \ln (2x) = 1 + \ln a$. *(C)*

141

4 Using the standard rules for differentiation and assuming the derivatives of $\sin \theta$ and $\cos \theta$, show that

$$\frac{d}{d\theta} (\cot \theta) = -\operatorname{cosec}^2 \theta \qquad \text{and} \qquad \frac{d}{d\theta} (\operatorname{cosec} \theta) = -\operatorname{cosec} \theta \cot \theta.$$

(3 marks)

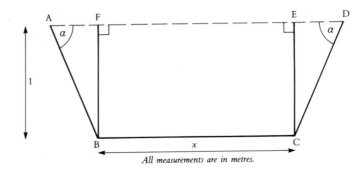

All measurements are in metres.

In the figure above, ABCD represents the cross-section of a rainwater channel. BC and AD are horizontal and the total length of AB, BC and CD is 4. Given that $BF = CE = 1$, $BC = x$ and $\widehat{DAB} = \widehat{ADC} = \alpha$, show that

$$x + 2 \operatorname{cosec} \alpha = 4$$

and deduce that $\alpha \geqslant \frac{1}{6}\pi$. *(4 marks)*

Show further that the cross-sectional area Y is given by

$$Y = 4 + \cot \alpha - 2 \operatorname{cosec} \alpha.$$

Hence show that the maximum value of Y occurs when $\alpha = \frac{1}{3}\pi$ and find this maximum value correct to 2 decimal places. *(6 marks)*

Show that if ABCD is bent into the shape of a semi-circle, an even larger cross-sectional area can be obtained. *(2 marks)*

(W)

5 The points A, B, C have position vectors $\mathbf{i}$, $\mathbf{j}$, $\mathbf{k}$ relative to an origin O, where $\mathbf{i}$, $\mathbf{j}$, $\mathbf{k}$ form a set of mutually perpendicular unit vectors. The points P, Q, lie on AB, BC respectively and are such that

$$\frac{AP}{PB} = \frac{BQ}{QC} = \frac{\lambda}{1 - \lambda}$$

where $0 \leqslant \lambda \leqslant 1$.

a) Write down the position vectors of P and Q and deduce a vector expression for $\mathbf{PQ}$ in terms of $\mathbf{i}$, $\mathbf{j}$, $\mathbf{k}$ and λ. Show that

$$|\mathbf{PQ}| = \sqrt{6(\lambda - \tfrac{1}{2})^2 + \tfrac{1}{2}}$$

and hence find the minimum and maximum lengths of PQ as λ varies. *(5 marks)*

b) The angles between **PQ** and **OA**, **PQ** and **OB** and **PQ** and **OC** are denoted by α, β, γ respectively. Obtain expressions for $\cos \alpha$, $\cos \beta$, $\cos \gamma$ in terms of λ and show that

$$\cos \alpha + \cos \beta + \cos \gamma = 0 \qquad \textit{(5 marks)}$$

c) Given that $\lambda = \frac{2}{3}$, find the position vector of the point of intersection of AQ and CP. *(5 marks)*

(W)

6

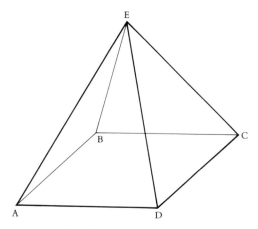

The figure shows a square-based pyramid in which

$$AB = BC = CD = DA = AE = BE = CE = DE.$$

Find the angles between

a) the lines AE and CD,

b) the line AE and the plane ABCD,

c) the planes ABE and ABCD.

(W)

7 A curve has equation $y = \ln\left(x^2 + \dfrac{1}{x^2}\right)$. Find $\dfrac{dy}{dx}$. Determine the nature and coordinates of any stationary points. *(7 marks)*

(O&C/MEI)

8 The points A and B have coordinates $(2, 3, -1)$ and $(5, -2, 2)$ respectively. Calculate the acute angle between AB and the line with equation

$$\mathbf{r} = \begin{pmatrix} 2 \\ 3 \\ -1 \end{pmatrix} + t \begin{pmatrix} 1 \\ -2 \\ -2 \end{pmatrix},$$

giving your answer correct to the nearest degree. *(4 marks)*

(C)

9 A curve is defined by the parametric equations

$$x = \cos^3 t, \qquad y = \sin^3 t, \qquad 0 < t < \frac{\pi}{4}.$$

Show that the equation of the normal to the curve at the point P $(\cos^3 t, \sin^3 t)$ is

$$x \cos t - y \sin t = \cos^4 t - \sin^4 t.$$

Prove that

$$\cos^4 t - \sin^4 t = \cos 2t.$$

The normal at P meets the x-axis at A and the y-axis at B. Express the length of AB as an integer multiple of cot 2t. *(10 marks)*
(JMB)

10 a) i) Show that

$$\frac{d}{dx}\left(\frac{x^3}{1+x^4}\right) = \frac{x^2(3-x^4)}{(1+x^4)^2}.$$

ii) Find

$$\int_1^2 \frac{x^2(3-x^4)}{(1+x^4)^2}\, dx. \qquad \text{(4 marks)}$$

b) Use the trapezium rule with subdivisions at $x = 3$ and $x = 5$ to obtain an approximation to

$$\int_1^7 \frac{x^3}{1+x^4}\, dx,$$

giving your answer correct to three places of decimals. *(4 marks)*

By evaluating the integral exactly, show that the error in the approximation is about 4.1%. *(4 marks)*
(C)

Part 4: Answers

Part 1: Useful Facts and Exercises

Manipulative Algebra

EXERCISE 1a

1 $(2x + 1)(x + 3)$
2 $(2x - 1)(x + 3)$
3 $(5x - 2)(x - 4)$
4 $(2x + 1)(2x + 3)$
5 $(4x - 1)(4x - 3)$
6 $(6x + 5)(3x - 4)$
7 $(4a + 1)(4a - 3)$
8 $(5x - 7)(3x + 5)$
9 $(4x + 5)(2x - 3)$
10 $(3a + 5)(3a - 4)$
11 $2(x + 2)(x - 3)$
12 $3(x + 5)(x + 7)$
13 $3(1 + a)(3 + 4a)$
14 $4(x + 2)(3x - 1)$
15 $(a + 5b)(a - 5b)$
16 $3(4x + 1)(4x - 1)$
17 $4(2 + 5a)(2 - 5a)$
18 $3(2x + 1)(2x - 1)$
19 $(a + 1)(b + 1)$
20 $2(a - 1)(b + 1)$
21 $(4 - x)(3 - x)$
22 $(4 + x)(5 - 2x)$
23 $(3 + a)(7 - a)$
24 $(5 - b)(3 + b)$
25 $(a + b)(2a - 1)$
26 $(m + 2n)(n - 3)$
27 $(2p + 3)(q + 2)$
28 $(2p - 1)(q + 2)$
29 $(x + y)(a - b)$
30 $(a - b)(x - 2y)$
31 $(a + b)(a - b + 3)$
32 $(a - b)(3 - a - b)$
33 $(a + b)(a + b + 2)$
34 $(a + b)(a - b - 2)$
35 $(a + 1)(a^2 - a + 1)$
36 $(2x + 1)(4x^2 - 2x + 1)$
37 $a(a - 1)(a^2 + a + 1)$
38 $(x^2 + 2y^2)(x^4 - 2x^2y^2 + 4y^4)$
39 $3(1 - 10x)(1 + 10x + 100x^2)$
40 $2(2a - 3b)(4a^2 + 6ab + 9b^2)$

EXERCISE 1b

1 $\dfrac{ac}{b^2}$
2 1
3 $4xz^3$
4 $\dfrac{1}{20}$
5 $2a$
6 $\dfrac{3}{y}$
7 $a - 3$
8 $\dfrac{2a + 3}{3a + 5}$
9 $\dfrac{4a - 3}{3a - 4}$
10 -1
11 $\dfrac{3}{x + 6}$
12 $a - 3$
13 $\dfrac{3}{a + 1}$
14 $\dfrac{x}{x + 5}$
15 $\dfrac{2a}{(a + b)(a - b)}$
16 $\dfrac{x - y - 1}{(x + y)(x - y)}$

17 $\dfrac{(5 + 3x)(5 - 3x)}{3(4 - 3x)}$
18 $\dfrac{x}{(1 + x)(1 - x)}$
19 $\dfrac{1}{a + 2}$
20 $\dfrac{a + 6}{(a + 3)(a - 3)}$
21 1
22 $\dfrac{7}{12(x + y)}$
23 $\dfrac{2}{(1 + x)(1 - x)}$
24 $-\dfrac{1}{(x + 1)}$
25 $\dfrac{3}{(x - 1)(x + 2)}$
26 $\dfrac{2}{(x + 1)(x + 2)}$
27 $\dfrac{2(x - 2)}{(x + 1)(x + 3)}$
28 $\dfrac{x + 18}{(x - 3)(2x + 1)}$
29 $-\dfrac{2(2a + 5)}{(a + 2)(a + 3)}$
30 $\dfrac{2a^2 - ab + 9b^2}{(a + 3b)(a - 2b)}$

EXERCISE 1c

1 $b = \dfrac{P}{2} - a$
2 $i = \sqrt{\dfrac{E}{R}}$
3 $a = 2s - b - c$
4 $a = \dfrac{2A}{h} - b$
5 $a = \dfrac{y^2}{4x}$
6 $a = \dfrac{v - u}{t}$
7 $R = \dfrac{P - L}{2\pi}$
8 $h = \dfrac{3V}{\pi r^2}$
9 $\theta = \dfrac{2A}{r^2}$
10 $p = \dfrac{Wb}{a}$
11 $m = \dfrac{y - c}{x}$
12 $A = 180 - 2C$
13 $a = \dfrac{v^2 - u^2}{2s}$
14 $u = \sqrt{v^2 - 2as}$
15 $c = \dfrac{ab}{a + b}$
16 $b = \dfrac{ac}{a - c}$
17 $c = \sqrt{b^2 - a^2}$
18 $r = \sqrt{\dfrac{3V}{\pi h}}$

19 $m = \dfrac{T}{g - a}$ 20 $c = \sqrt{a^2 - b^2}$

5 $\dfrac{y^2}{4a}$ 6 $\dfrac{ad + b(y - c)}{d}$

21 $v = \dfrac{2S}{t} - u$ 22 $r = \sqrt{R^2 - \dfrac{A}{\pi}}$

7 $\dfrac{a - yb}{y - 1}$ 8 $\dfrac{a}{y + b}$

23 $h = \dfrac{A}{2\pi r} - r$ 24 $y = -\dfrac{(Ax + C)}{B}$

9 $\dfrac{ab}{a + b}$ 10 $a + b$

25 $r = \dfrac{S - a}{S}$ 26 $v = \sqrt{\dfrac{2E}{m} + u^2}$

11 $a + b$ 12 $a - b$
13 $x = 4a, y = a$ 14 $x = 3b, y = -b$
15 $x = 3a, y = a$ 16 $x = 6p, y = 4p$

27 $r = 100\dfrac{(A - P)}{P}$ 28 $r = \dfrac{nE - CR}{Cn}$

17 $x = a + b, y = a - b$
18 $x = ab, y = 3b$

29 $C = \frac{5}{9}(F - 32)$ 30 $g = \dfrac{4\pi^2 L}{T^2}$

21 $v_1 = \frac{1}{10}(u_1 + 9u_2), v_2 = \frac{1}{5}(3u_1 + 2u_2)$
22 $v_1 = \frac{1}{3}(2 - e)u, v_2 = \frac{2}{3}(1 + e)u$
23 $y = c$

EXERCISE 1d

24 $a = \frac{1}{4}g, T = \frac{15}{4}g$

1 $-\dfrac{(by + c)}{a}$ 2 $\dfrac{y - c}{m}$

27 $R = \dfrac{9W}{5}, S = \dfrac{3W}{5}$

3 $\dfrac{a^2 + b^2 - c^2}{2ab}$ 4 $\dfrac{a(b - y)}{b}$

28 $a = \dfrac{mg \sin \alpha - Mg \sin \beta}{M + m}$

Quadratic Equations 2

EXERCISE 2

2 $p = -5, q = 6$
3 $p = -4, q = 1$
4 $x^2 + x - 12 = 0$
5 $x^2 - 10x + 18 = 0$
6 $4x^2 + 4x - 3 = 0$
7 a) real b) real c) imaginary
 d) imaginary
8 b), c) and d)
9 a) 7 b) 5 c) 39 d) $\frac{7}{5}$ e) $\frac{39}{5}$
10 a) 56 b) $3\frac{1}{2}$ c) -416 d) -416
11 a) $\frac{7}{4}$ b) $-\frac{73}{12}$ c) $\frac{73}{9}$ d) $\dfrac{\sqrt{97}}{3}$

 e) $-\dfrac{7\sqrt{97}}{9}$

12 a) $qx^2 + px + 1 = 0$
 b) $x^2 - (p^2 - 2q)x + q^2 = 0$
 c) $qx^2 - (p^2 - 2q)x + q = 0$
 d) $x^2 + (p - 2)x + 1 - p + q = 0$
 e) $x^2 + 4q - p^2 = 0$
 f) $q^2x^2 - (p^2 - 2q)x + 1 = 0$
13 $x^2 - 2x - 4 = 0$
14 $25x^2 - 29x + 4 = 0$
15 $x^2 + 7x + 2 = 0$
16 $k = -2$
17 $c = 4$
18 8 or -8

19 $\alpha + \beta = -\dfrac{b}{a}, \quad \alpha\beta = \dfrac{c}{a}$

 $a^3x^2 + abcx + c^3 = 0$

Partial Fractions 3

EXERCISE 3

1 $A = 5, B = 6$
2 $A = 1, B = -1$
3 $A = 1, B = -3, C = -9$
4 $A = 1, B = 1$
5 $A = 1, B = 2, C = -3$
6 $A = 1, B = 2$
7 $A = 4, B = -1, C = 2$
8 $A = 1, B = -1$

9 $\dfrac{2}{x} + \dfrac{3}{x - 4}$

10 $\dfrac{5}{x - 3} - \dfrac{2}{x - 4}$

11 $\dfrac{2}{x - 3} + \dfrac{4}{x - 4} - \dfrac{3}{x - 2}$

12 $\dfrac{2}{x-1} + \dfrac{3}{x+2} - \dfrac{1}{x-3}$

13 $-\dfrac{1}{8(x+1)} + \dfrac{1}{8(x-1)} - \dfrac{1}{4(x-1)^2}$
$+ \dfrac{1}{2(x-1)^3}$

14 $\dfrac{5}{2(x-1)} - \dfrac{1}{2(x+1)} - \dfrac{2}{x+2}$

15 $\dfrac{2x}{x^2+1} - \dfrac{2}{x+2}$

16 $-\dfrac{1}{3(x+1)} + \dfrac{x+1}{3(x^2-x+1)}$

17 $\dfrac{2}{x-3} - \dfrac{4x+1}{2x^2+1}$

18 $1 - \dfrac{3}{x+2} + \dfrac{1}{x+3}$

19 $2 + \dfrac{1}{x-1} - \dfrac{3}{x+2}$

20 $-\dfrac{1}{x+2} + \dfrac{3}{x-2} - \dfrac{2}{x+1}$

21 $\dfrac{1}{2a(x-a)} - \dfrac{1}{2a(x+a)}$

22 $1 - \dfrac{1}{x+1} + \dfrac{1}{x-1}$

23 $x + 1 + \dfrac{1}{x-2} - \dfrac{3}{x-4}$

24 $\dfrac{2}{x-4} - \dfrac{2}{x+1} - \dfrac{10}{(x+1)^2}$

25 $\dfrac{3}{x+1} + \dfrac{2}{(x+1)^2} + \dfrac{1}{(x+1)^3} - \dfrac{1}{x-2}$

Surds and Surd Equations 4

EXERCISE 4

1 a) $\sqrt{8}$ b) $\sqrt{27}$ c) $\sqrt{20}$ d) $\sqrt{63}$
2 a) $2\sqrt{3}$ b) $3\sqrt{2}$ c) $5\sqrt{2}$ d) $3\sqrt{5}$
3 a) $3\sqrt{2}$ b) $10\sqrt{2}$ c) 6 d) 6

4 a) $\dfrac{\sqrt{2}}{2}$ b) $\sqrt{3}$ c) $\dfrac{\sqrt{2}}{2}$ d) $5\sqrt{5}$

  e) $\sqrt{2} - 1$ f) $\sqrt{3} + 1$ g) $\frac{1}{2}(\sqrt{5} - 1)$
  h) $4\sqrt{2} + 2$ i) $2\sqrt{3} - 1$ j) $\sqrt{3}$
  k) -4
5 a) $2\sqrt{2} - 2$ b) $4\sqrt{3}$ c) 2 d) 1
  e) 26 f) $19 - 6\sqrt{2}$ g) $\sqrt{6}$
  h) $15(1 - \sqrt{2})$ i) $17 - 4\sqrt{15}$

6 a) $\sqrt{2} + 1$ b) $\sqrt{3} + \sqrt{2}$ c) $2 + \sqrt{3}$

  d) $2 + \dfrac{\sqrt{2}}{2}$ e) $4(\sqrt{2} - 1)$ f) $\frac{7}{10}\sqrt{5}$

  g) $3 + 2\sqrt{2}$ h) $5 - 2\sqrt{3}$ i) $\dfrac{3\sqrt{5} - 7}{4}$

7 3   8 1, 3
9 0, 4   10 $3, \frac{11}{9}$
11 3   12 $\frac{2}{9}, 2$
13 7   14 5
15 2, 6   16 -3

Arithmetic Progressions 5

EXERCISE 5

1 a) 25, b) 440   2 $-8, -20$
3 22   4 $12, 13\frac{1}{2}, 15, 16\frac{1}{2}$
5 $-4, 2$ 6 27, 19
7 2, 11, 600   8 10 000
9 10 200   10 29, 11 571

12 50   13 $8, \frac{1}{2}$
14 1.6, 0.2, 70   15 -374
16 $\log_e a, \log_e r, 9$ 17 20
18 11, 14, 17   19 $3a - b, 3a + b$
20 $2a - b, 3a - 2b, 4a - 3b$

Geometric Progressions 6

EXERCISE 6

1 $-3.665 \times 10^{11} \ [-\frac{1}{3}(2^{40})]$
2 $2, \frac{5}{2}, 157\frac{1}{2}$
3 £2048, £4095   4 $\pm\frac{1}{2}, 16$
5 3, 6, 12   6 8, 16, 32   7 11
8 a) $-\frac{1}{4}, \frac{1}{3}$ b) $13\frac{1}{2}$

10 b) $2\log_3 x$ 11 15 m
12 £$2^{28} - 1$ 13 2^{63}
14 a) £$1.08P$ b) £$1.08^2 P$ c) £$1.08^n P$,
   during the 10th year.
15 18   16 4, 8, 16

The Remainder Theorem 7

EXERCISE 7

1. a) 4 b) -9 c) 59 d) 1 e) 46
   f) -204

3. a) $(x-1)(x+2)(x-3)$
   b) $(x-1)(x+2)(x+4)$
   c) $(x+2)(x-3)(x+4)$
   d) $(x+1)(x+2)(x-4)$
   e) $(x+1)(x+2)(x+3)$
   f) $(x+2)(x-2)(x^2+4)$
   g) $(x-3)(x^2+3x+9)$
   h) $(x+1)^2(x-1)$
   i) $(x-3)(x^2+1)$
   j) $(x-1)(x-2)(x^2+1)$

   k) $(2x+1)(x+1)(x-3)$
   l) $(2x-1)(x+2)(x+3)$

4. $a = -6$
5. $b = 11$
6. $a = -10$
7. $b = 3$
8. $a = -3, b = -11, x-3$
9. $a = -4, b = -1,$
   $(x-1)(x-2)(x+2)(x-3)$
10. $a = -3, -4$
11. $a = -2$, other root is -1
12. $-2, 3, -3$
13. $2, 3, -4$
14. $a = -4, b = -1$, other roots are 1 and 3

Radian Measure 8

EXERCISE 8

1. a) $90°$ b) $45°$ c) $60°$ d) $30°$
   e) $135°$ f) $120°$ g) $360°$ h) $630°$

2. a) $\frac{\pi}{4}$ b) $\frac{\pi}{3}$ c) $\frac{2\pi}{3}$ d) $\frac{3\pi}{4}$
   e) $\frac{3\pi}{2}$ f) $\frac{5\pi}{3}$ g) $\frac{8\pi}{3}$ h) $\frac{10\pi}{3}$
   i) 4π

3. a) 0.8^c b) 1.2^c

4. $\frac{5^c}{8}$, 20 cm^2
5. 15 cm^2
6. 1.25^c
7. a) 1.350^c b) 10.8 cm c) 43.2 cm^2
8. 126.8 cm^2
9. a) $\frac{1}{2}r^2 \sin\theta$ b) $\frac{1}{2}r^2\theta$ c) $\frac{1}{2}r^2(\theta - \sin\theta)$
10. 123 cm^2

The Sine Rule and the Cosine Rule 9

EXERCISE 9

1. $a = 8.599$ cm $b = 11.16$ cm
2. $A = 60.6°, C = 34.9°,$
   area $\triangle ABC = 23.07$ cm^2
3. a) $AC = 122.5$ m, $BC = 36.6$ m
   b) 1585 m^2
4. a) 298 m b) 41.47 m
5. $c = 6.391$ cm, $A = 45.1°$
6. $41.48°$
7. a) $74.9°$ b) 27.90 cm^2

8. $A = 46.2°$ or $133.8°,$
   $a = 10.95$ cm or 24.86 cm
9. $82\frac{1}{2}°$, 11.96 cm
10. 2.746 km on a bearing of $329.2°$
11. a) 794 m on a bearing of $084.4°$
    b) 331 m
    c) i) 110 000 m^2 ii) 11 hectares
12. a) $22.8°$ b) $55.2°$
    c) 12.89 km on a bearing of $037.2°$

Trigonometric Identities 10

EXERCISE 10

5. a) $\frac{16}{65}$ b) $\frac{33}{65}$
6. a) $-\frac{36}{85}$ b) $\frac{13}{85}$ c) $\frac{84}{13}$
7. a) $\frac{4-6\sqrt{2}}{65}$ b) $\frac{8\sqrt{2}-3}{65}$
8. a) $\frac{2+\sqrt{3}}{2\sqrt{5}}$ b) $-\frac{1}{\sqrt{5}}$

9. $\tan x = \dfrac{\cos\alpha + \sin\beta}{\sin\alpha + \cos\beta}$

28. $\dfrac{x^2}{a^2} + \dfrac{y^2}{b^2} = 1$

29. $\dfrac{x^2}{a^2} - \dfrac{y^2}{b^2} = 1$

Trigonometric Equations 11

EXERCISE 11

1 30°, 150°
2 36.9°, 323.1°
3 45°, 225°
4 68.2°, 248.2°
5 109.3°, 250.7°
6 228.6°, 311.4°
7 15°, 75°, 195°, 255°
8 60°, 120°, 240°, 300°
9 75.4°, 104.6°, 195.4°, 224.6°, 315.4°, 344.6°
10 15°, 75°, 135°, 195°, 255°, 315°
11 30°, 150°, 221.8°, 318.2°
12 60°, 120°, 240°, 300°
13 63.4°, 161.6°, 243.4°, 341.6°
14 109.5°, 120°, 240°, 250.5°

15 90°, 270°
16 −180°, −60°, 60°, 180°
17 −150°, −30°, 90°
18 −180°, −120°, −60°, 0, 60°, 120°, 180°
19 −133.3°, 133.3°
20 −116.6°, −26.6°, 63.4°, 153.4°
21 $\frac{\pi}{8}, \frac{5\pi}{8}, \frac{9\pi}{8}, \frac{13\pi}{8}$
22 $\frac{3\pi}{4}$
23 $\frac{\pi}{12}, \frac{17\pi}{12}$
24 $\frac{\pi}{8}, \frac{5\pi}{8}, \frac{9\pi}{8}, \frac{13\pi}{8}$
25 $\frac{\pi}{3}, \pi, \frac{5\pi}{3}$
26 $0, \frac{2\pi}{3}, \frac{4\pi}{3}, 2\pi$
27 127.5°
28 0.361ᶜ, 2.780ᶜ

The Equation $a \sin \theta \pm b \cos \theta = c$ 12

EXERCISE 12

1 $5 \sin (\theta + 53.1°)$
2 $\sqrt{13} \sin (\theta − 33.7°)$
3 $\sqrt{5} \cos (\theta − 26.6°)$
4 $\sqrt{41} \sin (\theta + 51.3°)$
5 $\sqrt{13} \cos (\theta − 56.3°)$
6 $\sqrt{5} \cos (\theta + 26.6°)$
7 $5 \sin (2\theta + 53.1°)$, max value 5, when $\theta = 18.5°$

8 $\sqrt{13} \cos (\theta + 33.7°)$, min value $−\sqrt{13}$, 146.3°
9 103.3°, 330.5°
10 53.2°, 180°
11 12.3°, 105.7°
12 $R = 2, \alpha = 30°, \theta = 30°$
13 max 13 when $\theta = 22.6°$, min −13 when $\theta = 202.6°$
14 60°
15 45°, 161.6°, 225°, 341.6°
16 130.2°, 342.4°

Trigonometry in Three Dimensions 13

EXERCISE 13

1 a) 45°   b) 35.3°   c) 54.7°
2 a) i) 73.7°   ii) 21.8°   iii) 21.8°
   iv) 26.6°   v) 84°
   b) BG = 5 cm, GD = $2\sqrt{5}$ cm, BD = $\sqrt{13}$ cm, $\hat{B} = 60.1°$, $\hat{D} = 75.6°$, $\hat{G} = 44.3°$
3 a) 54.7° $(\tan^{-1} \frac{10}{\sqrt{50}})$ b) 12.2 cm $(\sqrt{150})$

   c) 11.2 cm $(\sqrt{125})$ d) 63.4° $(\tan^{-1} 2)$
4 a) 64.9°   b) 71.7°
   c) 9.055 cm   d) 95.7°
5 a) 6.928 cm   b) 6.532 cm
   c) 54.7°   d) 70.5°
6 a) 7.461 cm   b) 68.8°
   c) 63.6°   d) 96.6°

The Straight Line 14

EXERCISE 14

1 5
2 13, $(1, \frac{1}{2})$
3 (7, 3)
4 (−1, 1)
5 (−3, −2)
6 above
7 $\sqrt{74}$
9 a) i) 63.4° $(\tan^{-1} 2)$ ii) 68 sq. units
   iii) (5, −3)
10 $5x + 2y − 13 = 0$, 29 sq. units
11 90 sq. units
12 a) $5x − 9y − 17 = 0$
    b) $x + 2y − 1 = 0$
    c) $3x − 2y − 11 = 0$

    d) $4x − 3y + 20 = 0$
    e) $4x − 3y − 12 = 0$
15 a) $y = 2x$
    b) $x + y = 10$
    c) $16x + 2y − 21 = 0$
    d) $x^2 + y^2 − 6x + 1 = 0$
17 2 : 3
18 a) $4x + y − 17 = 0$ b) N(4, 1)
    c) $\sqrt{153} = 3\sqrt{17}$ d) 25.5 sq. units
19 $(4\frac{1}{2}, 1)$
20 (5, 1)
21 (2, 2)

Indices and Logarithms — 15

EXERCISE 15

1 3	**2** 2	**3** 2	**4** 1.292
5 3.807	**6** 2.151	**7** 3	**8** 3
9 0		**10** $\frac{1}{2}$	

12 a) $\log 72$ b) $\log 2$ c) $4 \log 3$

13 a) 2   b) 3

14 $\dfrac{\log_c a}{\log_c b}$

22 1   **23** 3   **24** 1.322

Log–log and Other Straight-line Graphs — 16

EXERCISE 16

1 error pair: $X = 5.2$, $Y = 435$;
  $a = 5.0$, $n = 2.6$

2 $a = 1.8$, $n = 0.8$

3 $m = 1.6$, $N = 20$

4 $K = 0.7$

5 $a = 1.5$, $b = 400$, a) 1900   b) 13

6 $a = 5$, $b = 30$, 6.2, 3.162

7 $a = 3.2$, $n = 1.7$

8 a) $y = \ln a + b \ln x$
  b) $a = 20$, $b = 0.5$
  c) i) 4.04   ii) 2.72

Permutations and Combinations — 17

EXERCISE 17

1 24	**2** 40 320	**3** 30 240	
4 210	**5** 30 240	**6** 210	
7 66	**8** 330	**9** 15 504	

10 a) $\dfrac{7!}{3!}$ b) $\dfrac{9!}{6!}$ c) $\dfrac{12!}{8!}$ d) $\dfrac{7!}{5!\,2!}$

  e) $\dfrac{9!}{6!\,3!}$

11 a) 120   b) 42   c) 604 800

12 a) 10   b) 35   c) 28

13 8!   **14** $2 \times 9!$ **15** 120, 24, 6

16 $\dfrac{9!}{2!\,2!}$ **17** $\dfrac{10!}{3!\,3!\,2!}$ **18** 7! (5040)

19 $\dfrac{9!}{2}$ **20** 8!   **21** 192   **22** 220

23 $\dfrac{15!}{4!}$ (10 695)   **24** $2 \times \dfrac{20!}{10!\,10!}$

25 140   **26** 1680   **27** 560

28 45   **29** 128   **30** 3600

31 1024 (4^5) **32** 9

The Binomial Theorem — 18

EXERCISE 18

1 $1 + 6x + 12x^2 + 8x^3$

2 $1 + 12x + 54x^2 + 108x^3 + 81x^4$

3 $1 + 5x + 10x^2 + 10x^3 + 5x^4 + x^5$

4 $1 - 8x + 24x^2 - 32x^3 + 16x^4$

5 $1 - \dfrac{3}{2}x + \dfrac{3}{4}x^2 - \dfrac{x^3}{8}$

6 $x^3 + 3x + \dfrac{3}{x} + \dfrac{1}{x^3}$

7 $27x^3 - 27x^2 + 9x - 1$

8 $x^4 - 8x^2 + 24 - \dfrac{32}{x^2} + \dfrac{16}{x^4}$

9 $81 + 54x + \dfrac{27}{2}x^2 + \dfrac{3}{2}x^3 + \dfrac{x^4}{16}$

10 $1 - \dfrac{7}{2}x + \dfrac{21}{4}x^2 - \dfrac{35}{8}x^3$

11 $1 + 24x + 252x^2 + 1512x^3$

12 $1 - 25x + \dfrac{1125}{4}x^2 - 1875x^3$

13 $2^{10} + 5 \times 2^{10}x + 45 \times 2^8 x^2 + 120 \times 2^7 x^3$
  $(1024 + 5120x + 11\,520x^2 + 15\,360x^3)$

14 $3^{12} - 8 \times 3^{12}x + 88 \times 3^{11}x^2 - 1760 \times 3^9 x^3$

15 $2^9 + 3 \times 2^8 x + 4 \times 2^7 x^2 + \dfrac{7}{9} \times 2^8 x^3$
  $(512 + 768x + 512x^2 + \dfrac{1792}{9}x^3)$

16 $924x^6$

17 $2^6 \times 3^7 \times 77x^5$, $2^7 \times 3^6 \times 77x^6$

18 $1 - 13x + 70x^2 - 196x^3 + 280x^4$

19 $a = -6$, $b = 15$

20 $a = -18$, $b = \dfrac{297}{2}$

21 $a = \dfrac{1}{6}$, $b = -\dfrac{7}{12}$

22 $1 - 6x + 27x^2 - 80x^3$

23 -10 **24** $-11\,536$ **25** 1120

26 84 **29** $n = 23$ **30** 8

31 $1.000\,099\,98$ **32** $1.002\,991$

33 $1 - \dfrac{3}{2}x - \dfrac{9}{8}x^2 - \dfrac{27}{16}x^3$, $0.969\,54$

34 $1 - \dfrac{2}{3}x - \dfrac{4}{9}x^2 - \dfrac{40}{81}x^3$, $|x| < \frac{1}{2}$, $0.979\,59$

35 $18\,564$ **36** $12\,870$

Inequalities 19

EXERCISE 19

1 $(-\infty, 1) \cup (4, \infty)$

2 $(-3, 3)$ **3** $(-\frac{5}{2}, 3)$ **4** $(-2, 7)$

5 $(-\frac{3}{2}, \frac{3}{2})$ **6** $(-4, 3)$ **7** $(3, 4)$

8 $(-\infty, \frac{1}{3}) \cup (1\frac{3}{5}, \infty)$

9 $(-3, -\frac{5}{3}) \cup (1, \infty)$

10 $(-\infty, \frac{3}{4}) \cup (2\frac{1}{2}, \infty)$

11 $(-\infty, 0.62) \cup (1.62, \infty)$

12 $(-0.22, 0) \cup (2.22, \infty)$

13 $(-\frac{1}{3}, 7)$

14 $(-\infty, 1.27) \cup (4.73, \infty)$

15 $(-\frac{\pi}{4}, \frac{\pi}{4}) \cup (\frac{3\pi}{4}, \frac{5\pi}{4})$

16 $(-2, 4)$

Simultaneous Equations 20

EXERCISE 20a

1 $x = 2, y = 4$; $x = -4, y = -2$

2 $x = 3, y = 4$; $x = -4, y = -3$

3 $x = 1, y = 3$; $x = -\frac{4}{5}, y = -\frac{3}{5}$

4 $x = 2, y = 3$; $x = 7, y = -2$

5 $x = -1, y = 3$; $x = -\frac{17}{5}, y = -\frac{9}{5}$

6 $x = \frac{3}{5}, y = \frac{8}{5}$; $x = -\frac{1}{2}, y = \frac{1}{2}$

7 $x = 3, y = 9$; $x = -7, y = -1$

8 $x = 4, y = 13$; $x = -5, y = -5$

9 $x = -5, y = -1$; $x = -\frac{2}{3}, y = \frac{4}{9}$

10 $x = -\frac{1}{2}, y = \frac{1}{8}$; $x = -\frac{2}{3}, y = 0$

11 $x = 4, y = 11$; $x = -1, y = 1$

12 $x = 5, y = 21$; $x = -2, y = -7$

13 $x = \frac{1}{3}, y = 2\frac{1}{3}$; $x = -1\frac{1}{2}, y = 6$

14 $x = 1, y = 3\frac{1}{2}$; $x = -2, y = -4$

15 $x = 3, y = 4\frac{1}{3}$; $x = \frac{1}{3}, y = -\frac{1}{9}$

16 $(0, 5)$ and $(4, -3)$

17 $(2, 3)$ and $(-\frac{34}{12}, -\frac{21}{13})$

EXERCISE 20b

1 $x = 3, y = 2$; $x = 1, y = 6$

2 $x = -\frac{5}{2}, y = -\frac{5}{2}$, $x = 1, y = 1$;
$x = -1.79, y = 2.68$;
$x = 0.86, y = -1.29$

3 $x = \pm 2, y = \pm 1$; $x = \pm 3, y = \pm 2$

4 $x = 1, y = -2$; $x = -1, y = 2$;
$x = 2, y = -1$; $x = -2, y = 1$

5 $x = 3, y = 2$; $x = -3, y = -2$;
$x = 2, y = 3$; $x = -2, y = -3$

6 $x = 2, y = 3$; $x = -2, y = -3$;
$x = 3, y = 2$; $x = -3, y = -2$

7 $x = 2, y = 3$; $x = -2, y = -3$;
$x = 3\sqrt{\frac{3}{2}}, y = 2\sqrt{\frac{2}{3}}$;
$x = -3\sqrt{\frac{3}{2}}, y = -2\sqrt{\frac{2}{3}}$

8 $x = 3, y = -1$; $x = -1, y = 3$

9 $x = 3, y = 1$;
$x = -2 + \sqrt{5}, y = 3(\sqrt{5} + 2)$;
$x = -2 - \sqrt{5}, y = 3(2 - \sqrt{5})$

10 $(4, 2)$, $(-4, -2)$, $(2, 4)$, $(-2, -4)$

11 $(\sqrt{2}, 3\sqrt{2})$, $(\sqrt{2}, -3\sqrt{2})$, $(-\sqrt{2}, 3\sqrt{2})$, $(-\sqrt{2}, -3\sqrt{2})$

Differentiation

EXERCISE 22a

1. $4x^3 + 6x$
2. $5(x + 3)^4$
3. $14(2x + 4)^6$
4. $12x(x^2 - 3)^5$
5. $-\dfrac{5}{(5x + 2)^2}$
6. $\dfrac{6}{(3 - 2x)^4}$
7. $2x - \dfrac{3}{x^2}$
8. $1 + \dfrac{4}{x^2}$
9. $\dfrac{2}{\sqrt{4x + 2}}$
10. $-\dfrac{x}{\sqrt{5 - x^2}}$
11. $\dfrac{3}{2\sqrt{x}} + \dfrac{2}{x\sqrt{x}}$
12. $9\sqrt{x}$

18. $\dfrac{\cos x}{2\sqrt{\sin x}}$
19. $\sqrt{\sec 2x}\,\tan 2x$
20. $-\dfrac{1}{4}\operatorname{cosec}^2\tfrac{1}{2}x\sqrt{\tan\tfrac{1}{2}x}$
21. $\dfrac{\cos^2 x + 1}{2\sqrt{\cos^3 x}}$
22. $\sec x\left(1 + x\tan x\right)$
23. $-\dfrac{2x\sin x + \cos x}{2\sqrt{x^3}}$
24. $x\left(2\tan x + x\sec^2 x\right)$

EXERCISE 22b

1. $2x(10x + 1)(4x + 1)^2$
2. $(5 - 4x)(5 - x)^2$
3. $\dfrac{7x^2 + 2}{2\sqrt{x}}(63x^2 + 2)$
4. $\dfrac{1}{(1 + x)^2}$
5. $\dfrac{1 - x}{(1 + x)^3}$
6. $\dfrac{4(1 + x)}{(1 - x)^3}$
7. $\dfrac{2x}{(1 + x^2)^2}$
8. $\dfrac{4x}{(1 - x^2)^2}$
9. $\dfrac{1}{\sqrt{(1 + x^2)^3}}$
10. $\dfrac{3(2 - x)}{2\sqrt{(1 - x)^3}}$
11. $-\dfrac{(x + 2)}{2x^2\sqrt{x + 1}}$
12. $\dfrac{1}{\sqrt{(x - 1)}\,\sqrt{(x + 1)^3}}$

EXERCISE 22c

1. $-\dfrac{1}{2}\sin\dfrac{1}{2}x$
2. $2\sin 4x$
3. $-3\sin 6x$
4. $2\sec^2 2x$
5. $\dfrac{1}{2}\sec\dfrac{1}{2}x\tan\dfrac{1}{2}x$
6. $-9\operatorname{cosec} 3x\cot 3x$
7. $-\operatorname{cosec}^2\dfrac{1}{2}x$
8. $4\sec^2 2x\tan 2x$
9. $\sec^2\dfrac{1}{2}x\tan\dfrac{1}{2}x$
10. $\cos x\cos 2x - 2\sin x\sin 2x$
11. $\dfrac{2\cos\frac{1}{2}x\cos x + \sin\frac{1}{2}x\sin x}{2\cos^2\frac{1}{2}x}$
12. $\sin x\left(2\cos^2 x - \sin^2 x\right)$
13. $-\dfrac{1}{4}\cos\left(\dfrac{\pi}{2} - \dfrac{x}{4}\right)$
14. $\cos x$
15. $-\cos x\dfrac{(1 + \sin^2 x)}{\sin^2 x}$
16. $3\tan^2 x\sec^2 x$
17. $\sec^2 x - \operatorname{cosec}^2 x$

EXERCISE 22d

1. $\dfrac{1}{x}$
2. $\dfrac{4}{x}$
3. $-\dfrac{1}{x}$
4. $\dfrac{4}{4x + 1}$
5. $\cot x$
6. $-2\tan 2x$
7. $\operatorname{cosec} x$
8. $1 + \ln x$
9. $\dfrac{1}{2}\tan\dfrac{1}{2}x$
10. $3e^{3x}$
11. $4xe^{2x^2}$
12. $-4e^{-4x}$
13. $e^x(\cos x + \sin x)$
14. $-\dfrac{3}{e^{3x}}$
15. $-e^{-x}\left(\sin x + \cos x\right)$
16. $e^{-x}\left(\sec^2 x - \tan x\right)$
17. $e^x + e^{-x}$
18. $a^x(1 + x\ln a)$
19. $\dfrac{2}{(x + 1)(3x + 1)}$
20. $-\dfrac{4}{4x + 3}$
21. $\cot x + 2\tan 2x$
22. $\dfrac{1}{1 + x} + \cot x$
23. $\dfrac{2x}{x^2 - 1} + \sec x\operatorname{cosec} x$
24. $\tan x + \sec x\operatorname{cosec} x$

EXERCISE 22e

2. $\dfrac{4}{3}$
3. a) $-\dfrac{1}{12}xy^4$ b) $-\dfrac{2x}{3y^2}$ c) $\dfrac{x}{12y^2}$
4. $2, 5$
5. $\dfrac{13}{9}$

EXERCISE 22f

1 $\dfrac{3t}{2}$ 2 $-\dfrac{1}{t^2}$ 3 $-t$ 4 $-\tan\theta$

5 $\dfrac{b}{a}\csc\theta$ 6 $\dfrac{b(1+t^2)}{2at}$

7 $\dfrac{t(2+3t)}{1+2t}$, $\dfrac{2(3t^2+3t+1)}{(1+2t)^3}$

8 $\frac{1}{4}(3-10t)$, $\frac{3}{10}$

9 $\dfrac{t^2+1}{t^2-1}$, $-\dfrac{4t^3}{(t^2-1)^3}$

Rate of Change   23

EXERCISE 23

1 $\frac{3}{4}$, gradient of tangent at $(-3, 4)$

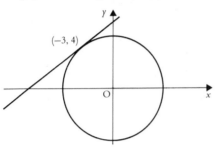

2 $225\ \text{cm}^3/\text{second}$ 3 $\frac{4}{3}\ \text{cm/second}$

4 a) $\dfrac{1}{\sqrt{2\pi}}$ cm/second (0.399),

   b) $\frac{1}{\pi}$ cm/second (0.318)

5 a) i) $5\ \text{m s}^{-1}$, $-12\ \text{m s}^{-2}$ ii) $-7\ \text{m s}^{-1}$, 0
      iii) $5\ \text{m s}^{-1}$, $12\ \text{m s}^{-2}$
   b) twice, when $t = 1$ and when $t = 5$

6 $-\frac{84}{55}$ cm/second $(-\frac{24}{5\pi})$. The height is
   decreasing while the radius increases.

7 v decreases by 0.71% (approx.)

8 6 cm, increasing at the rate of $\frac{6}{7}$ cm/second

9 a) 0.064 m/min   b) 0.192 m/min

10 0.322 cm/min

Stationary Values — Curve Sketching   24

EXERCISE 24

1 a) $(1, 0)$, $(4, 0)$, $(0, 4)$ b) min. $\left(\frac{5}{2}, -\frac{9}{4}\right)$

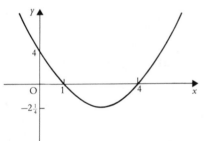

2 a) $(-6, 0)$, $(3, 0)$, $(0, 18)$
   b) max. $\left(-\frac{3}{2}, 20\frac{1}{4}\right)$

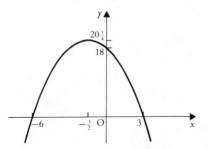

3 a) $(0, 0)$, $(1, 0)$, $(4, 0)$
   b) max. $(0.46, 0.88)$ min. $(2.87, -6.06)$

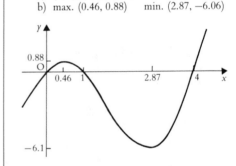

4 a) $(-4, 0)$, $(0, 0)$, $(3, 0)$
   b) min. $(-2.36, -21)$ max. $(1.7, 12.6)$

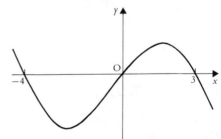

5 a) $(1, 0)$, $(2, 0)$, $(4, 0)$, $(0, -8)$
   b) max. $(1.45, 0.63)$ min. $(3.22, -1.95)$

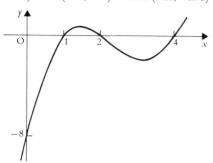

6 a) $(-1, 0)$, $(5, 0)$, $(0, -5)$
   b) min. $(2, -9)$

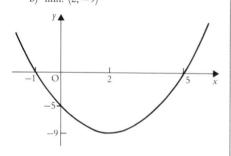

7

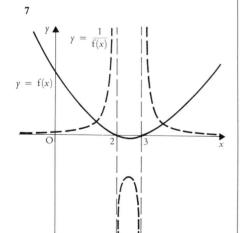

8 $a = 9$, $b = 15$, $c = 5$ min. $(5, -20)$

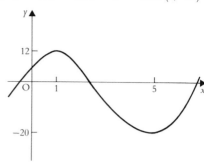

9 $a = -1$, $b = 3$, $c = 3$ min. at $x = -1$

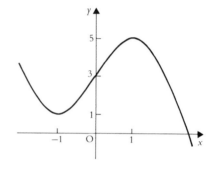

10 a) $(a, 0)$, $(2a, 0)$ b) $(h - a, k)$
    c) $(h + a, -k)$

11 a) A $(a, 0)$, B $(2a, 0)$
    b)

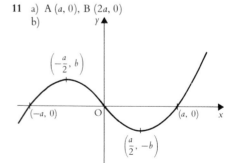

crosses x-axis at $(-a, 0)$, $(0, 0)$, and $(a, 0)$

max. at $\left(-\dfrac{a}{2}, b\right)$ min. at $\left(\dfrac{a}{2}, -b\right)$

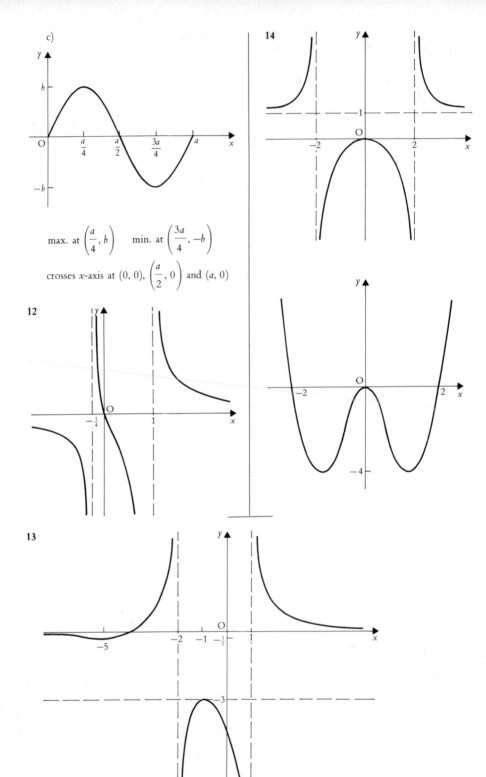

c)

max. at $\left(\dfrac{a}{4}, b\right)$ min. at $\left(\dfrac{3a}{4}, -b\right)$

crosses x-axis at $(0, 0)$, $\left(\dfrac{a}{2}, 0\right)$ and $(a, 0)$

14

12

13

15

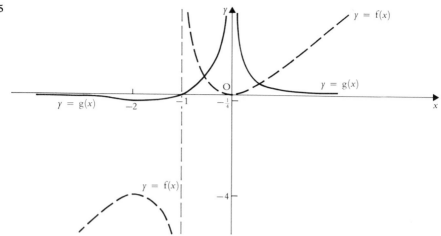

16

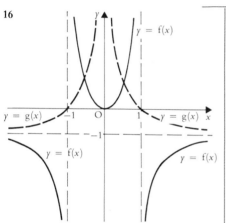

17 a) tangent $y = \frac{1}{2}$ normal $x = 1$

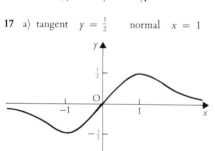

b) $y = \frac{1}{2}$

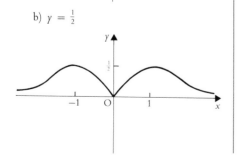

18

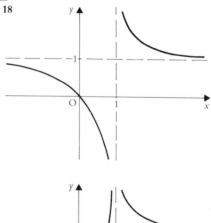

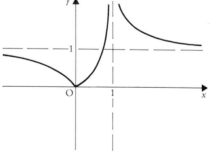

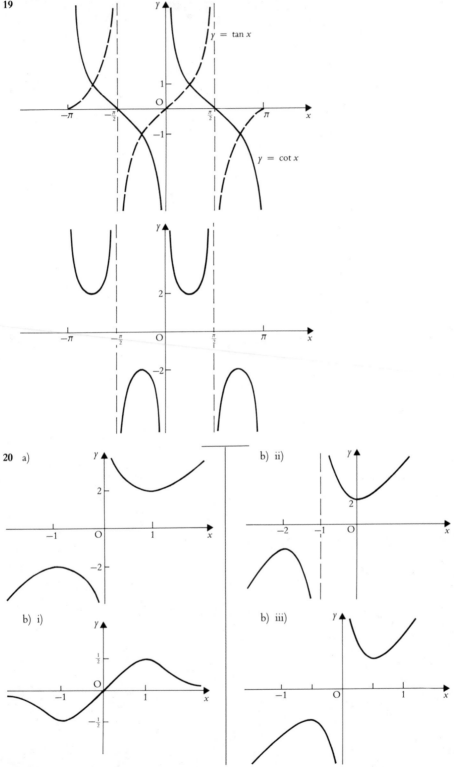

19

$y = \tan x$

$y = \cot x$

20 a)

b) ii)

b) i)

b) iii)

158

21

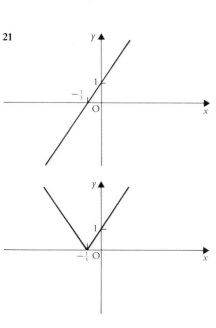

22

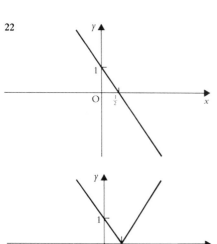

23

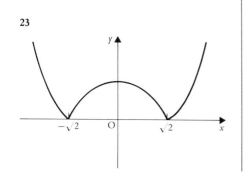

24

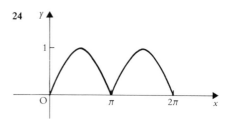

25

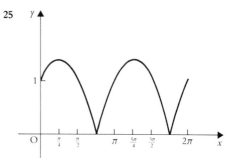

26

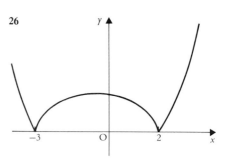

27

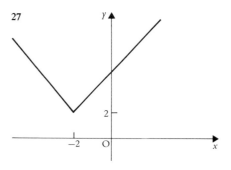

28

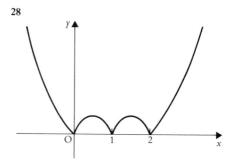

Problems Involving Maxima and Minima   25

EXERCISE 25

2   54 cm^3 3 $6 \text{ m} \times 3 \text{ m}$

4   $2 \text{ m} \times 2 \text{ m} \times 2 \text{ m}$

5   $l = 45 - 3x, \ 10, \ 1500$

8   $\dfrac{8000\pi}{27} \text{ cm}^3$

9   square, side $\dfrac{6}{4 + \pi}$ cm;

     circle, radius $\dfrac{3}{4 + \pi}$ cm

10   a) $2(a - x)$

     b) $\sqrt{2ax - a^2}, \ (a - x)\sqrt{2ax - a^2}$

11   $\dfrac{4\sqrt{3}}{9} \pi a^3$

12   $C = 25\sqrt{1600 + x^2} + 15(100 - x),$
     $x = 30, \ \pounds 2360$

The Circle   26

EXERCISE 26

1   a) $x^2 + y^2 = 9$
     b) $x^2 + y^2 - 4x - 6y + 9 = 0$
     c) $x^2 + y^2 - 8x + 4y + 4 = 0$
     d) $x^2 + y^2 + 6x + 4y - 12 = 0$
     e) $x^2 + y^2 - x + 6y + 7 = 0$

2   a) $(0, 0), 2$ b) $(2, 3), 5$ c) $(-3, 4), 1$
     d) $(-1, -5), 4$ e) $(0, 0), \frac{5}{2}$
     f) $(\frac{1}{2}, -\frac{1}{2}), 2$

3   a) $(-2, 3), 4$ b) $(\frac{5}{2}, -\frac{3}{2}), \dfrac{3\sqrt{2}}{2}$

     c) $(0, 0), \frac{5}{3}$ d) $(1, -2), \dfrac{6}{\sqrt{5}}$

4   a) outside   b) on   c) inside
     d) outside

5   $g = -2, f = -1, c = -11$

6   $x^2 + y^2 - x - 2y - 14 = 0$

7   centre $(1, 1)$, radius $\sqrt{5}$,
     equation $x^2 + y^2 - 2x - 2y - 3 = 0$

8   $x^2 + y^2 - 4x + 2y - 20 = 0, \ (2, -1), 5$

9   a) $4x + 3y - 25 = 0$
     b) $2x + y - 8 = 0$
     c) $4x - 3y + 30 = 0$
     d) $3x + 2y - 26 = 0$

10   $(3, 0)$ and $(0, 2)$

11   $(2, 2)$ and $(-2, 6), 4\sqrt{2}$

12   cuts at $(-1, 2)$ and $(2, 3)$

13   AB is a diameter, yes

14   a) i) centre $(1, 2)$ ii) radius $2\sqrt{2}$

15   $x^2 + y^2 - 8x + 4y - 5 = 0$,
     $x^2 + y^2 - 8x - 8y + 7 = 0$

16   $x^2 + y^2 - 2x + 4y - 8 = 0$

18   $2x - y - 5 = 0, \ x - 2y + 5 = 0$

19   $3x - y - 3 = 0, \ x - 3y + 7 = 0, \ 53°$

20   $3x + y - 9 = 0, \ 53°$

21   a) C_1: centre $(-1, -2)$, radius 5
        C_2: centre $(5, \frac{5}{2})$, radius $\frac{5}{2}$
     b) $7\frac{1}{2}$ c) $(3, 1)$

22   $(5, 5)$; tangents: $S_1 \ 3x + 4y - 35 = 0$,
     $S_2 \ 4x - 3y - 5 = 0$; normals:
     $S_1 \ 4x - 3y - 5 = 0, \ S_2 \ 3x + 4y - 35 = 0$

23   A $(a, 0)$, B $(0, a)$, C $(-a, 0)$, D $(0, -a)$;
     $x + y = a$,
     $x \sin\theta - y(1 + \cos\theta) + a \sin\theta = 0$,
     $y = \dfrac{a \sin\theta}{1 + \cos\theta}$

24   C_1: $(2, 0)$, radius 2; C_2: $(0, 2)$, radius 2;
     A $(\frac{4}{5}, \frac{8}{5})$, B $(\frac{8}{5}, \frac{16}{5})$, C $(2, 2)$

Further Coordinate Geometry   27

EXERCISE 27a

1   a) $x - y + 2 = 0$ b) $x + y - 6 = 0$

2   a) $x - y + a = 0$ b) $x + y - 3a = 0$

3   $y = \dfrac{2}{p + q}x + \dfrac{2apq}{p + q}$

5   $x = -3a$

8   a) $pq = -1$ b) $p = q$

9   $2y = x + 4$, Q $(1, -2), x + y + 1 = 0$,
     $71.6°$

10   P $(\frac{1}{2}, 2), x + 2y + 8 = 0$,
     $4x + 3y - 8 = 0$, T $(-2, -3)$, R $(2, 0)$

11 $y = \dfrac{x}{t} + at$, R $(2at^2, 2at)$

12 $pq = -1$

13 M $\left[\dfrac{a}{2}\left(p^2 + \dfrac{1}{p^2}\right), \; a\left(p - \dfrac{1}{p}\right) \right]$

EXERCISE 27b

1 a) $9x - y - 17 = 0$
   b) $x + 9y - 11 = 0$

2 $y = 12, y = -15$

3 $(-2, 0), (0, 0), (3, 0), y = 10x + 20$,
   $y = -6x, y = 15x - 45$

4 $-8, 8x + y + 10 = 0, x - 8y + 50 = 0$

5 $3x - 2y - 12 = 0, 2x + 3y + 5 = 0$

8 $\dfrac{x}{a} + \dfrac{y}{b} = \dfrac{1}{\sqrt{2}}$

11 $y = \dfrac{b(t^2 - 1)x}{2at} + \dfrac{a(1 - t^2) + 2bt}{1 + t^2}$

14 T $\left(\dfrac{2ctt_1}{t + t_1}, \dfrac{2c}{t + t_1}\right), \quad y = \dfrac{x}{K}$

Functions

28

EXERCISE 28

1 $\{0, 1, 4\}$

2 $\{-8, -2, 1, 4, 7, 13\}$

3 $\{-1, 3, 5\}$, 3

4 $\{-7, -3, 1, 5, 9, 13\}$

5 $[4, \infty)$ i.e. $y: y \in \mathbb{R}, y \geqslant 4$

6 $(-\infty, 4]$ i.e. $y: y \in \mathbb{R}, y \leqslant 4$

7 $[-4, 4], [0, 4]$ **8** 2   **9** -1

10 a) 19   b) 5   c) 27   d) 39
   e) 1083   f) $3 + 12x^2$
   g) $3(9 + 24x + 16x^2)$

11 a) -4 b) -23 c) 7   d) -55
   e) $37 + 25x - 10x^2$
   f) $-25 + 85x - 50x^2$

12 $\{-4, 2, \frac{4}{5}\}$ **13** a) g   b) h   c) f

14 a) odd   b) even   c) odd

15 $5x + 28, \dfrac{x - 28}{5}$

16 $f^{-1}(x) = \dfrac{5x + 1}{x} \; (x \neq 0)$

     $g^{-1}(x) = \dfrac{1 - 4x}{5x} \; (x \neq 0)$

17 Both domain and range of f^{-1} are $[0, \infty)$

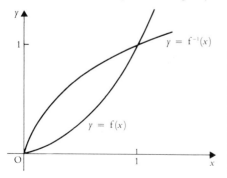

18 $f^{-1}(x) = -\frac{1}{3}\sqrt{1 - x^2}, \quad g^{-1}(x) = 2 \ln x$;
   for f^{-1} domain is $[0, 1]$ and range is
   $[-\frac{1}{3}, 0]$;
   for g^{-1} domain is $(0, \infty)$ and range is
   $(-\infty, \infty)$

19 $[0, 3]$, no, restricted domain $[0, 3]$;
   using restricted domain for f the domain
   for f^{-1} is $[0, 3]$ and the range is $[0, 3]$

20 a) $(-\infty, \infty), x \neq 2; (-2, \infty); [-5, 5]$

     b) $\ln\left|\dfrac{2(3x - 2)}{x - 2}\right|, (-\infty, \frac{2}{3}) \cup (2, \infty)$

     c) $f^{-1}(x) = \dfrac{2x}{x - 4} \; (-\infty, \infty), x \neq 4$;

       $g^{-1}(x) = e^x - 2, (-\infty, \infty)$

     d) $[-5, 0]$ or $[0, 5]$

21

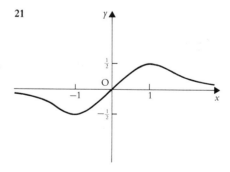

$k = 1$

For g^{-1} domain is $[-\frac{1}{2}, \frac{1}{2}]$ and range is
$[-1, 1]$

Integration

29

EXERCISE 29a

1 $\dfrac{5x^2}{2} - 4x + c$

2 $\dfrac{1}{12}(3x+2)^4 + c$

3 $-\dfrac{1}{3(3x-1)} + c$

4 $\dfrac{1}{2}x^2 - \dfrac{1}{x} + c$

5 $\dfrac{x^5}{5} - 2x^3 + 9x + c$

6 $x + \dfrac{4}{3}x\sqrt{x} + c$

7 $-\dfrac{2}{3}\sqrt{4-3x} + c$

8 $-\dfrac{5}{2x} + \dfrac{x^2}{4} + c$

9 $\dfrac{1}{10}(1+x^2)^5 + c$

10 $4\frac{1}{2}$ sq. units

11 each area is $\frac{1}{4}$ sq. unit

12 $\frac{1}{12}$ sq. unit

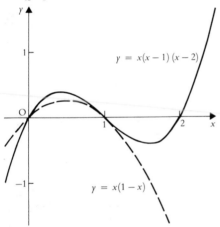

13 36 sq. units   **14** $57\frac{1}{6}$ sq. units

15 $y = x^3 - 6x^2 + 8x$, 4 sq. units

EXERCISE 29b

1 $-\dfrac{1}{3}\cos 3x$

2 $\dfrac{1}{4}\sin 4x$

3 $\dfrac{1}{2}\ln|\sec 2x|$

4 $\dfrac{1}{2}\tan 2x$

5 $\dfrac{x}{2} - \dfrac{\sin 4x}{8}$

6 $\dfrac{1}{2}(x + \sin x)$

7 $-\dfrac{3}{2}\cos\left(2x - \dfrac{\pi}{4}\right)$

8 $\dfrac{4}{3}\sin\left(3x + \dfrac{\pi}{6}\right)$

9 $\dfrac{1}{3}\sec 3x$

10 $\dfrac{x}{2} - \dfrac{\sin 6x}{12}$

11 $\dfrac{x}{2} - \dfrac{3}{4}\sin\dfrac{2x}{3}$

12 $\dfrac{1}{3}\sin^3 x$

13 $-\dfrac{1}{4}\cos^4 x$

14 $\dfrac{\sin^4 x}{4} - \dfrac{\sin^6 x}{6}$

15 $-\operatorname{cosec} x$

16 $\frac{1}{2}$

17 $\frac{1}{3}$

18 $\frac{1}{2}(\sqrt{2}-1)$

19 $\dfrac{3\sqrt{3}-1}{24}$

20 $\dfrac{1}{\sqrt{3}}$

21 $2\sqrt{2}$

22 $1 - \frac{\pi}{4}$

23 $\frac{\pi}{16}$

24 $\frac{11}{24}$

EXERCISE 29c

1 $\ln|x+1|$

2 $\dfrac{3}{2}\ln|2x+1|$

3 $\dfrac{1}{2}\ln|1+x^2|$

4 $-\dfrac{1}{3}\ln|1-x^3|$

5 $\ln|x^2+3x-4|$

6 $\dfrac{1}{2}\ln|x^2-4x+7|$

7 $\dfrac{1}{5}e^{5x}$

8 $-\dfrac{1}{3}e^{-3x}$

9 $-\dfrac{1}{2}e^{-2x}$

10 $\ln|1+e^x|$

11 $2(x - \ln|1+e^x|)$

12 $-\dfrac{2}{3\sqrt{e^{3x}}}$

13 $\dfrac{1}{5}\ln\frac{21}{11}$ (0.1293)

14 $\dfrac{e^5-1}{5e^5}$

15 $\dfrac{1}{3}(e^2-1)$

EXERCISE 29d

1

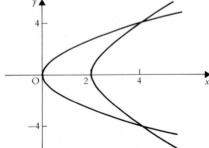

$10\frac{2}{3}$ sq. units, 16π cube units

2 $\frac{\pi}{2}$ cube units

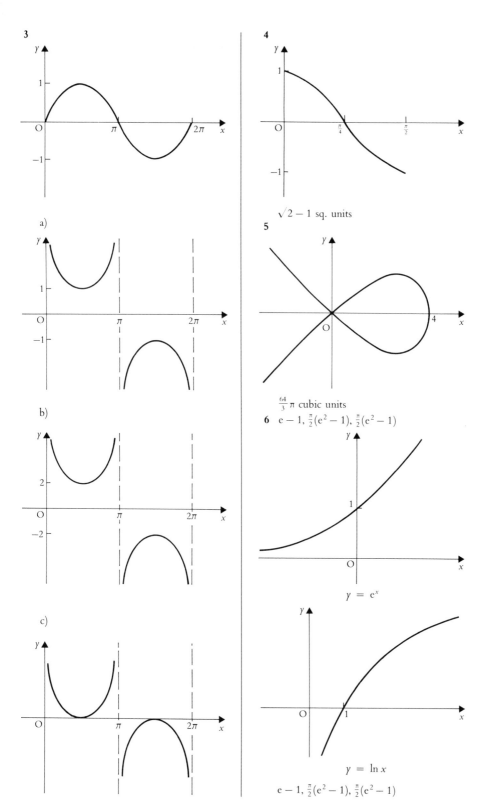

3

a)

b)

c)

4

$\sqrt{2} - 1$ sq. units

5

$\frac{64}{3}\pi$ cubic units

6 $\mathrm{e} - 1$, $\frac{\pi}{2}(\mathrm{e}^2 - 1)$, $\frac{\pi}{2}(\mathrm{e}^2 - 1)$

$y = \mathrm{e}^x$

$y = \ln x$

$\mathrm{e} - 1$, $\frac{\pi}{2}(\mathrm{e}^2 - 1)$, $\frac{\pi}{2}(\mathrm{e}^2 - 1)$

EXERCISE 29e

1 $\sin^{-1}\dfrac{x}{2}$ **2** $\sin^{-1}\dfrac{x}{3}$

3 $\frac{1}{2}\sin^{-1}\dfrac{2x}{3}$ **4** $\frac{1}{3}\tan^{-1}\dfrac{x}{3}$

5 $\frac{1}{4}\tan^{-1}\dfrac{x}{4}$ **6** $\frac{1}{2}\tan^{-1}\dfrac{x}{2}$

7 $\frac{1}{2}\tan^{-1}\left(\dfrac{x+1}{2}\right)$ **8** $\frac{1}{2}\sin^{-1}\left(\dfrac{x+1}{2}\right)$

9 $\frac{1}{2}\sin^{-1}2x$ **10** $\frac{\pi}{4}$ **11** $\frac{\pi}{2}$

12 $\frac{\pi}{4}$ **13** $\frac{\pi}{6}$

14 $\frac{1}{24}\left(\pi - 4\tan^{-1}\frac{3}{2}\right)(-0.0329)$

15 $\frac{3\pi}{4} - \frac{3}{2}\sin^{-1}\frac{2}{3}$ (1.262)

EXERCISE 29f

1 $\dfrac{1}{30(1-5x^2)^3}$ **2** $\frac{1}{36}(1+3x^2)^6$

3 $\frac{1}{3}\sqrt{(x^2+3)^3}$ **4** $\sqrt{x^2+3}$

5 $\frac{1}{5}\sqrt{(x^2+1)^5} - \frac{1}{3}\sqrt{(x^2+1)^3}$

6 $\tan\frac{1}{2}x$

7 $\tan\left(\dfrac{x}{2}-\dfrac{\pi}{4}\right)$ or $-\dfrac{2}{1+\tan\frac{1}{2}x}$

8 $\frac{\pi}{12}$ **9** $6+3\ln 7$ **10** $\frac{\pi}{4}$ **12** $\ln 3$

EXERCISE 29g

1 $\frac{1}{5}\ln\left|\dfrac{x-1}{x+4}\right|$ **2** $\ln\left|\dfrac{x^3}{(x+1)^2}\right|$

3 $\ln\left|\dfrac{x+1}{x-4}\right| - \dfrac{5}{x-4}$

4 $\frac{3}{2}\ln|x| - \frac{1}{4}\ln|x^2+2|$

5 $2\ln|x| - \ln|x^2+1| + \tan^{-1}x$

6 $\frac{2}{3}\ln\left|\dfrac{x-3}{x+3}\right|$

7 $x + \frac{1}{2}\ln\left|\dfrac{x-1}{x+1}\right|$ **8** $\frac{1}{6}\ln\left|\dfrac{x-1}{x+5}\right|$

9 $\ln|x-3| - \dfrac{6}{(x-3)} - \dfrac{4}{(x-3)^2}$

10 0.1178 **11** 0.4319 $\left(\ln 3 - \frac{2}{3}\right)$

12 -1.386 $\left(-\frac{1}{3}\ln 64\right)$

EXERCISE 29h

1 $(x-1)\,e^x$ **2** $x(\ln|x|-1)$

3 $\dfrac{x^2}{4}(2\ln|x|-1)$ **4** $\sin x - x\cos x$

5 $(x^2-2)\sin x + 2x\cos x$

6 $\frac{1}{25}e^{5x}(5x-1)$ **7** $-\frac{1}{4}(2x+1)\,e^{-2x}$

8 $x\sin^{-1}x + \sqrt{1-x^2}$

9 $\dfrac{(x^2+1)}{2}\tan^{-1}x - \dfrac{x}{2}$

10 $(x^2-2x+2)\,e^x$ **11** $\frac{1}{9}x^3(3\ln|x|-1)$

12 $\frac{1}{5}e^{2x}(2\cos x + \sin x)$

13 $e^x(x^3-3x^2+6x-6)$, 0.5634 $(6-2e)$

14 $\frac{1}{5}\tan^5 x - \frac{1}{3}\tan^3 x + \tan x - x$

EXERCISE 29i

1 2.021 **2** 1.724 **3** 4.209
4 0.9943 **5** 0.9432 **6** 0.2017

EXERCISE 29j

1 0.5000 **2** 1.187 **3** 0.7854
4 0.5346 **5** 0.2712 **6** 0.4142

Numerical Solutions to Equations   30

EXERCISE 30a

1 $x_4 = 7.6056$ **2** $x_4 = 0.3820$
(to 4 d.p.)   (to 4 d.p.)
3 2.7144 **4** 0.5825

EXERCISE 30b

1 1.60 **2** -2.90 **3** 1.24
4 0.24 **5** 0.51 **6** 0.45
7 1.93 **8** -3.183

Complex Numbers   31

EXERCISE 31

1 a) $5,\ 1+2i,\ 7-i,\ 1+i$
   b) $1-2i,\ -3+4i,\ 1+5i,\ -\frac{5}{13}-\frac{i}{13}$
     or $-\frac{1}{13}(5+i)$
   c) $7i,\ -i,\ -12,\ \frac{3}{4}$
   d) $1+2i,\ -7,\ -13+i,\ -\frac{11}{17}+\frac{7}{17}i$

2 a) $11-2i$ b) $-11-2i$
   c) $4i$ d) $\frac{13}{5}+\frac{11}{5}i$
   e) $\frac{9}{13}+\frac{6}{13}i$ f) $-\frac{2}{5}i$
   g) $\frac{1}{2}-\frac{3}{10}i$ h) $1+2i$

3 a) $-2 \pm i$
b) $1 \pm \sqrt{3}i$

c) $-3 \pm i$
d) $-\frac{5}{4} \pm \frac{\sqrt{7}i}{4}$

e) $\frac{3}{2} \pm i$
f) $1 \pm \frac{i}{2}$

4 a) $0, \pm i$
b) $1, 2 \pm i$
c) $1, -1 \pm i$
d) $2, 3 \pm 2i$

5 $-i, -2 \pm i$

6 $-3 - 2i, \pm 2i$

7 $\frac{ac + bd}{c^2 + d^2} + \frac{(bc - ad)}{c^2 + d^2}i$

8

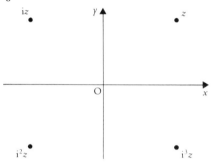

length of side of square is 2 units

	mod.	arg.
z	$\sqrt{2}$	$\frac{\pi}{4}$
iz	$\sqrt{2}$	$\frac{3\pi}{4}$
i^2z	$\sqrt{2}$	$-\frac{3\pi}{4}$
i^3z	$\sqrt{2}$	$-\frac{\pi}{4}$

9 a) $\sqrt{5}, 1.107^c$
b) $2, -1.047^c$

c) $\sqrt{7}, 0.714^c$
d) $\frac{2}{\sqrt{3}}, 0.524^c \left(\frac{\pi}{6}\right)$

e) $\sqrt{7}, 2.285^c$
f) $\sqrt{5}, -2.678^c$
g) $1, \alpha$
h) $\sec \alpha, \alpha$

i) $2\cos\frac{\alpha}{2}, -\frac{\alpha}{2}$

10 a) $4 + 5i$
b) $-4 - 5i$
c) $-5 + 4i$
d) $5 - 4i$

11 a) $\sqrt{2} + \sqrt{2}i$
b) $2\sqrt{3} + 2i$

c) $\frac{1}{2} - \frac{\sqrt{3}}{2}i$
d) $-2 - 2\sqrt{3}i$

e) $-2i$
f) $-\sqrt{2} + \sqrt{2}i$

12 a) $x^2 + y^2 = 4$
b) $x^2 + y^2 - 2x - 8 = 0$
c) $x^2 + y^2 + 2y - 8 = 0$
d) $x = 0$
e) $x = 3$
f) $4x + 2y - 3 = 0$

13

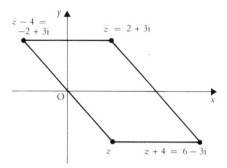

the four points form a parallelogram

14 $\pm (3 + 2i)$

15 a) $\frac{a}{a^2 + b^2} - \frac{b}{a^2 + b^2}i$
b) $a - bi$

c) $\frac{a}{a^2 + b^2} + \frac{b}{a^2 + b^2}i$

16 a) $\frac{1}{r}(\cos\theta - i\sin\theta)$

b) $r(\cos\theta - i\sin\theta)$

c) $\frac{1}{r}(\cos\theta + i\sin\theta)$

17 C: $7 + 3i$, D: $8 - i$

18 $\frac{13}{10} + \frac{13}{10}i$

19 R: $7 + 4i$, S: $8 - i$

20 a)

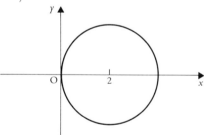

b)

165

c)

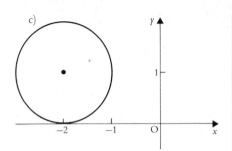

h)

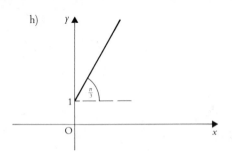

d)

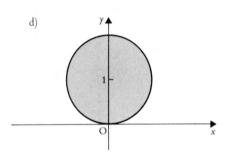

21 1, A: $-1 - 3i$, C: $3 + 3i$

22 a) $\pm (3 - i)$ b) $\pm (5 + 2i)$
c) $\pm (4 - 3i)$ d) $\pm (3 + 5i)$

23 a) $x^2 - 6x + 10 = 0$
b) $x^2 - 4x + 13 = 0$
c) $x^3 - 3x^2 + 4x - 2 = 0$
d) $x^3 - 6x^2 + 9x - 50 = 0$

24 a) $3(\cos 0 + i \sin 0)$
b) $4\left(\cos \frac{\pi}{2} + i \sin \frac{\pi}{2}\right)$
c) $\sqrt{2} \left(\cos \frac{\pi}{4} + i \sin \frac{\pi}{4}\right)$
d) $2\left(\cos \frac{\pi}{6} + i \sin \frac{\pi}{6}\right)$
e) $2\sqrt{3} \left[\cos \left(-\frac{\pi}{6}\right) + i \sin \left(-\frac{\pi}{6}\right)\right]$
f) $5\left[\cos \left(-2.214^c\right) + i \sin \left(-2.214^c\right)\right]$

e)

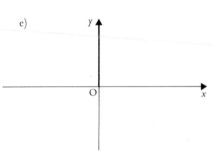

25 a)

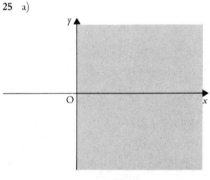

f)

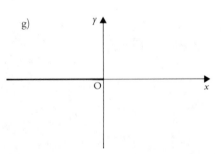

b)

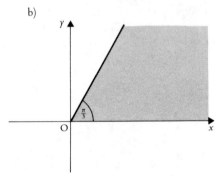

g)

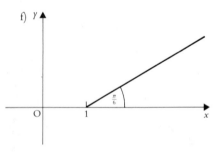

166

c)

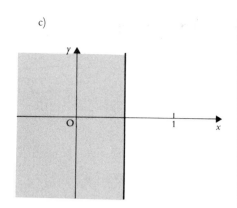

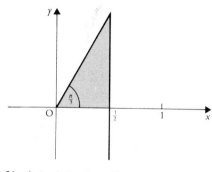

26 a) $1 + i, 2 + i$ b) $2 + i, -3 + 2i$
c) $1 + 2i, 3 + 3i$

Vectors 32

EXERCISE 32

1 $(4, 12, -6)$

2 $\overrightarrow{BC} = i - 3j - 2k$, $\overrightarrow{PQ} = \frac{1}{2}i - \frac{3}{2}j - k$
$2\overrightarrow{PQ} = \overrightarrow{BC}$, i.e. PQ is parallel to BC
and equal to half of it

3 $(\frac{2}{3}, -\frac{2}{3}, \frac{1}{3})$

4 $\dfrac{x-1}{2} = \dfrac{y-2}{1} = \dfrac{z+3}{3}$

5 yes   **6** no

7 $r = 3i + 4j - 2k + \lambda(4i + 3j + 2k)$;
$(7, 7, 0)$

8 $4i + 2j + 5k$ **10** $33.5°$, $(\cos^{-1}\frac{5}{6})$

12 $r = i - 2k + \lambda(-4i + j + 3k)$;
$-\dfrac{4}{\sqrt{26}}, \dfrac{1}{\sqrt{26}}, \dfrac{3}{\sqrt{26}}$

13 $73.2°$ $(\cos^{-1}\frac{11}{38})$ **14** 4

Differential Equations 33

EXERCISE 33

1 $x^2 - y^2 = A$ **2** $y = Ax$

3 $y = ax^4 + c$ **4** $4\ln x = 3\ln ky$

5 $y = A(x + 4)$ **6** $y = Ae^{\frac{1}{2}x^2}$

7 $y + 2 = c(x - 3)$

8 $y^2 = 2\sqrt{x^2 - 1} + A$

9 $y\ln\left|\dfrac{x+1}{Kx}\right| = 1$

10 $y = A(x^2 + 1)$

11 $\ln\dfrac{y-1}{y+1} = e^{2x} + A$

12 $v = (v + 1)A e^u$

13 $y = A\ln x$

14 $e^x = \ln\tan y + c$

15 $\ln y = -\dfrac{1}{x}\ln x - \dfrac{1}{x} + C$

167

Part 2: Revision Papers 1–10

Revision Paper $\qquad$ 1

1 a) $\frac{24}{25}$ b) $\frac{7}{25}$ c) $\frac{24}{7}$

3 43.6°, 17.8 cm²   **4** 32

5 $2x - y - 3 = 0$ **6** 0.0966 cm/min

7 $(2, 3), (-8, -7)$ **8** 8

9 $(-1, 4), (-\frac{8}{5}, -\frac{1}{5}), x + 2y - 7 = 0$, $2x - y + 6 = 0$

10 $x + p^2y - 2ap = 0$

11 $(x + 1)^2 + 2^2$, 4 when $x = -1$

Revision Paper $\qquad$ 2

1 $0, -1$ **2** 2, 4, 8 or 8, 4, 2

3 $\dfrac{1}{\sqrt{3}}$

4 $(\alpha + \beta)(\alpha^2 - \alpha\beta + \beta^2)$
   a) 8   b) 6   c) 32   d) $22\frac{1}{2}$

5 radius 8, $(-2, -3)$, 6

6 $a = -1, b = -9, (x - 1)(x + 3)(x - 3)$

7 a)

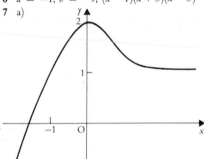

b)

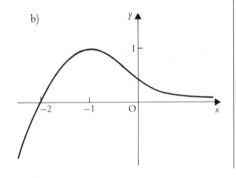

c)

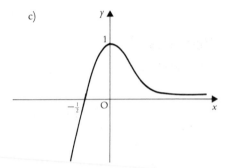

8 0.2727   **9** $(3, 1, 0)$

10 a) $e^{3x} \dfrac{(3 \sin 2x - 2 \cos 2x)}{\sin^2 2x}$ b) $\ln \dfrac{x}{x + 1}$

11 $a = 6, b = 9, c = 6$, max. 10 at $x = 1$

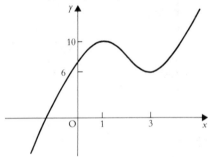

12 a) when $t = 3$ b) when $t = 1$ and 3
   c) -6 m s^{-2}, 6 m s^{-2} d) $t = 2, -3$ m s^{-1}

Revision Paper $\qquad$ 3

1

x	4	-4	3	-3
y	3	-3	4	-4

2 a) $-\dfrac{24}{25}$ b) $-\dfrac{7}{25}$ c) $\dfrac{24}{7}$

3 $1 - x + \dfrac{2}{3}x^2 - \dfrac{10}{27}x^3, |x| < 3$

5 44   **6** 2, 93   **7** -3

8 $-\dfrac{198}{125} - \dfrac{1089}{125}i$ **10** 2 radians, 114.6°

Revision Paper 4

1 $a^2(2b^2 - a^2)$

2 a) $-\dfrac{4\sqrt{5}}{9}$ b) $\dfrac{1}{4\sqrt{5}}$

4 $92.6°$, $22.3\ \text{cm}^2$

5 $(-\infty, -3) \cup (1\tfrac{1}{2}, \infty)$

6 $1 + 3x + \dfrac{5}{2}x^2 + \dfrac{x^3}{2}, \ |x| < \tfrac{1}{2}$

9 a) B $(1, 0)$, D $(5, 0)$ b) 8 sq. units

10 $1240\ \text{cm}^3$ (to 3 s.f.)

11 $\dfrac{1}{a}$, $y = \dfrac{1}{e}x$

12 a) 1.32   b) 88 cm

Revision Paper 5

1 $60°$, $70.5°$, $289.5°$, $300°$

5 $a = 2$, $b = -5$, -5 when $x = -2$

6 $x = 2$, $y = 2$; $x = -1$, $y = -4$

7 $1 - 8x + 44x^2 - 168x^3$

8 $x^2 + y^2 = 1$

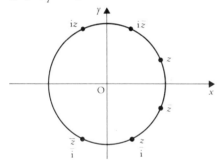

9 a) $p = 2$, $q = 4$
   b) GP $S_n = 2^{n+1} - 2$,
      AP $S_n = n(n + 1)$
   c) If $n = 6$, GP $S_6 = 126$,
      AP $S_6 = 42$

10 $3x - 4y + 28 = 0$, they have a common
    tangent at the point $(-4, 4)$

11 $\dfrac{1}{x - 2} + \dfrac{2}{x^2 + 4}$, $\ln(x - 2) + \tan^{-1}\tfrac{1}{2}x$

12 a) i) $\tfrac{1}{5}$ ii) $(3, -3)$ iii) $5x + y - 12 = 0$
    b) $x - 5y + 8 = 0$ c) $(2, 2)$
    d) $2\sqrt{13}$ e) $(2, 2)$
    f) $a = 2$, $b = 2$, $r = 2\sqrt{13}$

Revision Paper 6

1 $x = 5$, $y = 2$; $x = -2$, $y = -5$

2 a) $9u$ b) u^2,   1 or 1.631

4 0.7007

5 a) $\tfrac{11}{64}x^5$ b) $14 - (x - 3)^2$, 14 when $x = 3$

7 $x^2 - y^2 + x^2y^2 - 2y - 1 = 0$

9 $y = \dfrac{x}{t} + at$, $y = -tx - \dfrac{a}{t}$

10 $8x^2 + 8y^2 - 20x + 8 = 0$, radius $\tfrac{3}{4}$, the
    complex number $\tfrac{5}{4} + 0i$ represents the
    centre of this circle

11 a) $\dfrac{15}{4}x^2 + \dfrac{112}{x}\ \text{cm}^2$ b) $68.22\ \text{cm}^2$

12 $\tfrac{\pi}{3} \times x^2(30 - x)$, $\dfrac{1}{3\pi}\ \text{cm/min}$, 26.2 min

Revision Paper 7

2 a) $\frac{4}{7}\sqrt{7}$ b) $3\sqrt{7}$ c) 8   3 $-1, 2$

4 $\dfrac{1}{1+x} - \dfrac{1}{2x+1}$ 6 48.79

7 $(3, -5), (-1, 11)$; $4x - y - 17 = 0$, $4x - y + 15 = 0$

8 1.15^c

9

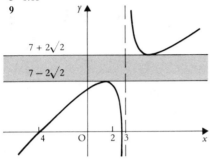

10 a) 2 or 3   b) 2 or -4 c) -2 or 6

11 a) 22 140   b) -820 c) 11 480
    d) 19 670

12 a) $\ln|x| - \frac{1}{2}\ln|x^2 + 1|$ b) $2e^{\frac{1}{2}x}(x - 2)$

Revision Paper 8

1 a) 1.89   b) $\frac{2}{3}$ 2 1.465

3 1.349 cm/s

4 $A = (s - x)\sqrt{2sx - s^2}$, $x = \frac{2}{3}s$

5 $a = 1, b = \frac{3}{2}, c = -\frac{21}{8}$ 7 0.805

8 8 acres   9 $x - 4y + 8 = 0, 41°$

10 $-1 + \frac{1}{2}i, 3 + \frac{1}{2}i$

11 a) $7\frac{1}{2}$ sq. units

b) $P_1 (2, 4), Q_1 (8, 1), R_1 (9, 3)$
c) $P_2 (2, 4), Q_2 (-1, -2), R_2 (-3, -1)$
d) $P_3 (14, -2), Q_3 (8, 1), R_3 (9, 3)$

12 $\dfrac{1}{x - 4} + \dfrac{2}{x + 3} - \dfrac{3}{(x + 3)^2}$,

$\log|x - 4| + 2\log|x + 3| + \dfrac{3}{x + 3} + c$

Revision Paper 9

1 $2(1 \pm \sqrt{2})$ $(-0.83$ and $4.83)$,
   $\dfrac{1 \pm \sqrt{17}}{8}$ $(-0.39$ and $0.64)$

2 $x = 3, y = 5; x = -2, y = 0$

4 $\sin x + x \cos x$ 5 1.768

6 $3, \frac{3}{4}, \frac{3}{16}, \frac{3}{64}$

8 $a = 2, 4x + 3y - 30 = 0$,
   $x + 3y - 12 = 0$

10 $z = \frac{1}{5} - \frac{7}{5}i, r = \sqrt{2}, \theta = -1.429^c$

	mod.	arg.
$-z$	$\sqrt{2}$	1.713
$\dfrac{1}{z}$	$\dfrac{1}{\sqrt{2}}$	1.429

	mod.	arg.
z^2	2	-2.858
$\bar{z}$	$\sqrt{2}$	1.429
$\dfrac{1}{\bar{z}}$	$\dfrac{1}{\sqrt{2}}$	-1.429

11 $\overrightarrow{YX} = 6i - 3j + 5k, \overrightarrow{YZ} = 8i + j + k,$
    $\overrightarrow{ZX} = -2i - 4j + 4k; XY = \sqrt{70},$
    $ZY = \sqrt{66}, ZX = 6; \overrightarrow{YX}.\overrightarrow{YZ} = 50,$
    $X\hat{Y}Z = 42.6°$

12 $A = 100x - 2x^2, x = 25,$
    area $= 50 \times 25 \text{ m}^2 = 1250 \text{ m}^2$,
    yes—if erected as a semicircle

Revision Paper

1 $x = 5, y = 6; x = -1, y = 2$

2 $43.3°, 316.7°$

4 $\tan(A + B + C) =$

$$\frac{\tan A + \tan B + \tan C - \tan A \tan B \tan C}{1 - \tan A \tan B - \tan B \tan C - \tan C \tan A}$$

5 $a = 25, b = -90$

7

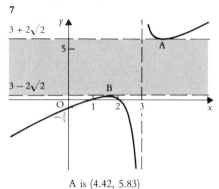

A is $(4.42, 5.83)$
B is $(1.59, 0.18)$

8

x	2	-2	$\dfrac{11\sqrt{3}}{3}$	$-\dfrac{11\sqrt{3}}{3}$
y	$\dfrac{5}{3}$	$-\dfrac{5}{3}$	$-3\sqrt{3}$	$3\sqrt{3}$

9 $\dfrac{n}{6}(n + 1)(2n + 1)$

11 $27.6 \text{ cm}^3, 4.01 \text{ cm/s}$

12 b) i) $\frac{1}{2}\tan^{-1} 2x$

ii) $\ln(x^2 + 1) + 3\tan^{-1} x$

Part 3: Past A-level Questions, Papers 1–15

Paper 1

1 $107.6°$, $252.4°$

2 a) $k \in [-3, 4\frac{1}{2}]$ b) $k \in (-\frac{1}{2}, \frac{1}{2})$

3 a)

x	27	$\frac{1}{9}$
y	$\frac{1}{3}$	$3\sqrt{3}$

   b) 100

4 a) $r = 1.030$ b) 72 679

5 a) $fg: x \mapsto -\ln x$, $f^{-1}: x \mapsto e^x$
   i) fg is a reflection of f in the x-axis
   ii) f^{-1} is a reflection of f in the line
       $y = x$
   b) Since one value of y gives two values
      of x, hf is not one–one
   c) $\phi^{-1}(x): x \mapsto e^{-\sqrt{x}}$

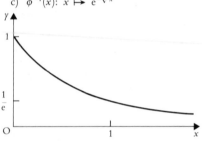

6 a) $79.9°$ b) 13.4 cm   c) $36.7°$

7

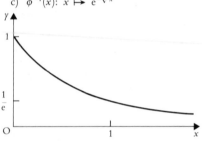

8 max. at $(\frac{3\pi}{2}, 4\frac{1}{2})$, min. at $(\frac{\pi}{2}, -3\frac{1}{2})$;
   inflexion at $(-2\pi, \frac{1}{2})$, $(-\pi, \frac{1}{2})$, $(0, \frac{1}{2})$, $(\pi, \frac{1}{2})$
   and $(2\pi, \frac{1}{2})$; crosses x-axis at -5.76^c,
   -3.67^c, 0.52^c, 2.62^c

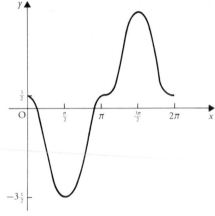

9 a) 3.79 b) 1.55 c) 3.359

10 $a = 2$, $b = 9$, 0.107 $(\frac{1}{3} \tan^{-1} \frac{1}{3})$

Paper 2

1 a) $\frac{1}{8}$
   b) $p = q^{\frac{1}{4}} + q^{\frac{3}{4}}$

2 a) 3 cm
   b) $92°$

3 a) $x^2 + y^2 - 6x - 10y + 9 = 0$, $(0, 9)$
   b) $(-1, 2)$ and $(7, 8)$

4 $f^{-1}: x \mapsto \left(\dfrac{x - 3}{4}\right)^{\frac{1}{3}}$ $(x \in \mathbb{R})$
   f^{-1} is the reflection of f in the line $y = x$
   and vice versa

5 b) $t = -1$ c) $\sqrt{68}$

6 centre $(1, 0)$, radius 1

7 a) $V = 3x^2d$, $A = 8xd + 6x^2$

8 b) i) $\tan 4\theta = \dfrac{4 \tan \theta \left(1 - \tan^2 \theta\right)}{1 - 6 \tan^2 \theta + \tan^4 \theta}$

9 a) 3.79   **b)** 1.55   **c)** 3.359

$$\int e^{2x} \cos 3x \, dx$$
$$= \tfrac{1}{13} e^{2x} \left(2 \cos 3x + 3 \sin 3x\right)$$
$$\int e^{2x} \sin 3x \, dx$$
$$= \tfrac{1}{13} e^{2x} \left(2 \sin 3x - 3 \cos 3x\right)$$

b) $\frac{11}{12}$

10 a) $\frac{1}{3}(2\sqrt{2} - 1)$

b) $\dfrac{2}{x - 1} - \dfrac{4x}{2x^2 + 1}$, $\ln \frac{36}{19}$

Paper   3

2 a) 3

3 0.16

4 a) $r^2\theta$ **c)** 1.166

5 a) A′ (4, 10), B′ (7, 4), C′ (5, 3)
   b) A″ (1, 1), B″ (7, 4), C″ (6, 6)

6 $f(x) = \dfrac{\pi}{2} - x - \dfrac{x^3}{6} - \dfrac{3}{40}x^5$

7 a) $74 + 70 \cos \left(\theta - 36.87°\right)$
   c) max. value 12, $\theta = 36.87°$;
     min. value 10.67, $\theta = 98.2°$

8 $\ln \left(1 + x\right) = x - \tfrac{1}{2}x^2 + \tfrac{1}{3}x^3 - \tfrac{1}{4}x^4$
   $e^{-x} = 1 - x + \tfrac{1}{2}x^2 + \tfrac{1}{6}x^3 + \tfrac{1}{24}x^4$
   $G(x) = 1 - x + \tfrac{1}{2}x^2 - \tfrac{1}{6}x^3 - \tfrac{1}{12}x^4$

9 a) $\tfrac{\pi}{2}\left(e^{2a} - e^{-2a}\right)$
   b) i) $\frac{1}{110}$ **ii)** $\left(x^2 + 1\right) \tan^{-1} x - x$

10

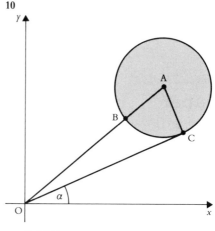

7, 0.644
least value $\alpha = 0.339$

Paper   4

1 a) $(2x - y)(2x - 3y)$, $x = \tfrac{1}{2}$, $y = -1$
   b) $x = -1$ or 2

2 -4, 0.59, 3.41

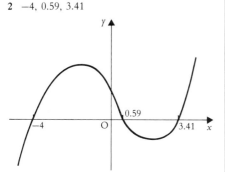

f(x) < 0 for $(-\infty, 4) \cup (0.59, 3.41)$ or
$x < -4$ and $0.59 < x < 3.41$

3 $R = 5$, $\alpha = 0.927^c$ $\left(\tan^{-1} \tfrac{4}{3}\right)$
   a) 2.1^c, 6.1^c **b)** $\theta = 0.93^c$, 0.31^c, 2.4^c,
     4.5^c A (0, 1), B (−2.21, 5), C $(0.927, \tfrac{5}{6})$

4 a) 59.0°   **b)** 53.1°   **c)** 111.1°

5 a) $z_1 = 2 + 3i$, $z_2 = 3 - 4i$
   b) $a = \dfrac{x^2 + y^2 - 6x - 8y}{(x - 6)^2 + y^2}$, centre (3, 4),
     radius 5

6 a)

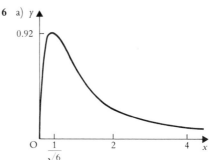

7 $p^2 y + x = 2cp$
   $T\left(\dfrac{2cpq}{p + q}, \dfrac{2c}{p + q}\right)$, $2xy = kx + hy$

8 a) 24 years   **b)** $A = 10\,000$, $B = 0.80$

9 a) $(-1, \frac{1}{2})$, $(0, 0)$, $(1, \frac{1}{2})$

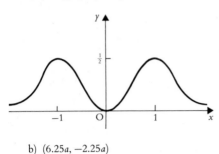

b) $(6.25a, -2.25a)$

10 a) $x\tan^{-1}x - \frac{1}{2}\ln|1 + x^2|$, $\frac{\pi}{4}$

b)

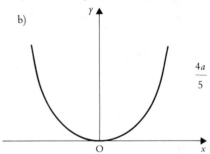

$\frac{4a}{5}$

Paper 5

1 $\sin 40° : \sin 60° : \sin 80°$; 0.258

2 a) i) $\dfrac{4}{1-x}$

　　ii) $1 + 4x + 4x^2 + 4x^3 + 4x^4 + 4x^5$

　　iii) $1 + 4x + 4x^2 + 4x^3 = 1 + \dfrac{4x}{1-x}$

$$= \dfrac{1 + 3x}{1 - x}$$

3 a) -2, $2y = x$　　b) $y = x + 1$

　　c) $x^2 + y^2 - 5x - 7y - 22 = 0$, $Q\left(\frac{44}{5}, \frac{22}{5}\right)$

4 $\frac{3}{4}\mathbf{a} + \frac{1}{4}\mathbf{b}$

6 a) $4x - 12y + 21 = 0$　　b) $-\frac{1}{4}$

　　c) $\frac{5}{7}$

7 $k = -\frac{1}{24}$

8 $-\frac{3}{2}\sin t$, $\frac{7}{8}a$

9 a) $-(3x + \frac{5}{2}x^2 + 3x^3 + \frac{17}{4}x^4 + \frac{33}{5}x^5)$, -0.0159

　　b) $a = 1$, $b = -1$

10 c) $\frac{1}{35}(x^2 - 3)^{\frac{5}{2}}(5x^2 + 6)$

Paper 6

1 $a = 2$, $b = 1$, $c = 3$, $(-1, 3)$

2 $(-2, -1) \cup (2, 3)$

3 $a = 4$, $b = -9$; $-1, 2, -3, 7$; $1, -2, 3, -7$

4 a) 1.998　　b) $\theta = n\frac{\pi}{2} + (-1)^n\frac{\pi}{12}$

5 0.66　　　　　**6** $\approx 14°$

7

Interval ...	A	B	C	D	E
Signs of $f(x)$	$-$	$+$	$+$	$+$	$-$
$f'(x)$	$+$	$+$	$-$	$-$	$-$
$f''(x)$	$-$	$-$	$-$	$+$	$-$

8

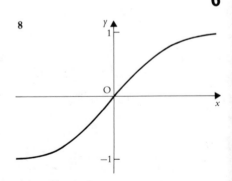

0.828 $(2(\sqrt{2} - 1))$, 1.348 $(\frac{\pi}{2}(4 - \pi))$

10 a) $t = 0$ or -4　　b) $72.5°$

Paper 7

1 a) 10   b) $(2, 2\frac{1}{2})$ c) 0 or $-\frac{2}{3}$

2 a) $1 + \frac{1}{5}x - \frac{2}{25}x^2 + \frac{6}{125}x^3$

$\quad 1 + \frac{1}{5}x - \frac{2}{25}x^2 + \frac{4}{125}x^3$

3 a)

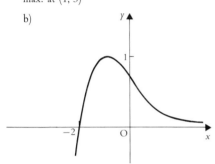

max. at $(1, 3)$

b)

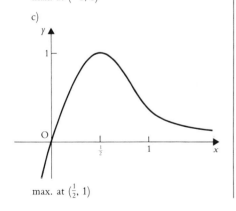

max. at $(-1, 1)$

c)

max. at $(\frac{1}{2}, 1)$

4 D $(2, 5)$ 5 a) $20°, 140°$ b) 0.44

6 min. $(1, 0)$; max. $(5, -8)$

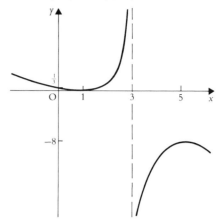

area $= 0.122$ sq. units

7 $\phi = 2\pi \sin\theta$, $\tan\theta = \sqrt{2}$,

$\quad$ area of S $= \dfrac{\sqrt{6}\pi R^2}{3}$

8 $a = 1.6$, $b = 2.6$

9 $y = 2x - 4$, N $(4, 4)$ a) $\frac{3}{2}$ b) $\frac{3}{4}$

10 a) i) $\ln|1 + x|$ ii) $x + \ln|1 + x|$

$\quad$ iii) $e^x(x - 1)$ b) π

Paper 8

1 a) $k \in (0, 12)$

$\quad$ b) $x^2 - (k^2 - 9k + 4)x + 2k^2 + 9k + 4 = 0$

$\quad k = 3$, double root is -7

2 $f(x) = 4x^2 - 2x + 1$ 3 1.557

4 $\dfrac{dV}{dt} = -kV$ a) £4495   b) 57 months

5 2.6^c

6 a) $\dfrac{\sqrt{3}a}{2}$ b) $\dfrac{\sqrt{5}a}{2}$

$\quad$ c) $50.8°$ $\left(\tan^{-1}\sqrt{\dfrac{3}{2}}\right)$ e) $\dfrac{2a}{\sqrt{5}}$, no

7 a)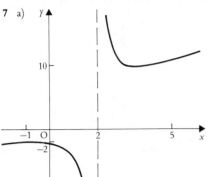

b)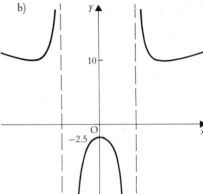

9 a) i) 55°   **ii)** 71°   **iii)** 35°

b) $\mathbf{AC} = 6\mathbf{i} + 8\mathbf{k}$, $|\mathbf{AB}| = 13$,

$|\mathbf{AC}| = 10$, $\mathbf{AB}.\mathbf{AC} = 114$

$\mathbf{OD} = (6\lambda - 2)\mathbf{i} + 3\mathbf{j} + (8\lambda - 7)\mathbf{k}$

10 0.911   **a)** yes   **b)** no

Paper

9

2 15°, 75°, 90°, 135°

3 2.35

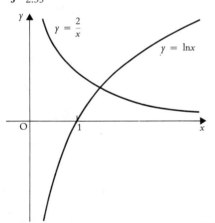

4 $Q\left(\frac{12}{5}, \frac{4}{5}\right)$

$R\left(-\frac{12}{5}, \frac{36}{5}\right)$

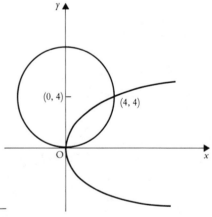

5

$x = 0, y = \sqrt{2}$

$x = \frac{\pi}{4}, y = 1$

$x = \frac{\pi}{2}, y = \sqrt{2}$

Volume generated is 2π

6 $y = \dfrac{3}{16}x^2 + \dfrac{1}{x^2}$

7 a) 298

b) $S_n = 8[1 - (-\tfrac{1}{2})^n]$, $S = 8$, 13

9 $f \circ g: x \mapsto e^{2x} - 2$, domain $(-\infty, \infty)$

$g^{-1}: x \mapsto \ln x$, domain $(0, \infty)$

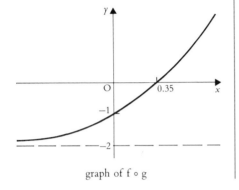

graph of f ∘ g

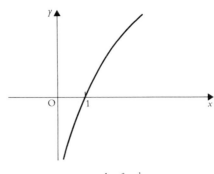

graph of g^{-1}

10 a) $\dfrac{6 - \sqrt{3}\pi}{12}$

Paper 10

1 $\theta = \dfrac{2n\pi}{3} \pm \dfrac{\pi}{6}$, $\alpha = 19.3°$

2 $\ln 2$

3 b) $h(x) = \dfrac{x^2 - 5}{2}$

4 a) $b = -0.25$, $\ln a = 2.06$

b) i) $(-\infty, 2] \cup [6, \infty)$ ii) $1, \dfrac{1}{2\sqrt{2}}$

5 radius 5, centre $(2, -3)$, $\pm \dfrac{(-6 - k)}{5}$

$k = 19$ or -31

6 a) $\dfrac{2}{p + q}$ b) $\dfrac{1}{p}$ c) $y + px = 2ap + ap^3$

$a(p^2 + pq + q^2 + 2)$, $-apq(p + q)$

7 0.18

8 a) $2\mathbf{i} + \mathbf{j} - 2\mathbf{k}$, b) 3, c) $\frac{5}{21}$,
d) 10.2 sq. units, -3, $\sqrt{17}$

9 a)

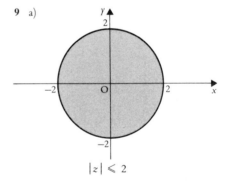

b)

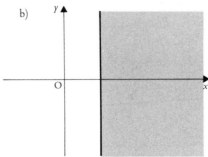

$\text{Re}(z) \geqslant 1$

177

c)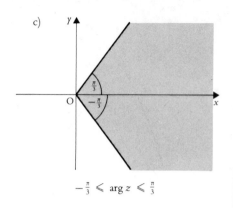

$$-\tfrac{\pi}{3} \leqslant \arg z \leqslant \tfrac{\pi}{3}$$

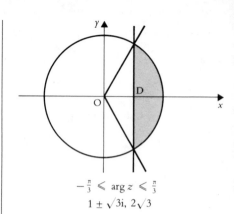

$$-\tfrac{\pi}{3} \leqslant \arg z \leqslant \tfrac{\pi}{3}$$
$$1 \pm \sqrt{3}i, \; 2\sqrt{3}$$

10 i) 65 cm   ii) 45°, 21°

Paper   11

2 $a = \dfrac{1-b}{2}, c = \dfrac{2-d}{2}, c = -b, b = -\tfrac{3}{5}$

4 $\mathbf{r} = 5\mathbf{i} + 3\mathbf{j} + 7\mathbf{k}$
$\mathbf{r} = (12 + 2\lambda)\mathbf{i} + 5\mathbf{j} + (6-\lambda)\mathbf{k}$
position vector of c is $6\mathbf{i} + 5\mathbf{j} + 9\mathbf{k}$

5 $\tfrac{1}{2} + \tfrac{3}{2}i, \; \tfrac{3}{2} + \tfrac{3}{2}i$

6 b) $4\sqrt{A}$

7 a) 1.443   b) $\dfrac{1}{\ln 2}$

8 $2, t = 0, \pi, 2\pi, 3\pi, \ldots$

9

$y = |x-1|$

$y = x^2 - 1$

$(-\infty, -2) \cup (1, \infty)$

10 a) 0.105, 0.317
b) $\tfrac{1}{2} e^{2x}(x^2 - x + \tfrac{3}{2})$ or $\tfrac{1}{4}(2x^2 - 2x + 3)e^{2x}$
c) $\tfrac{1}{2} \sec^{-1} \dfrac{x+2}{2}$

Paper   12

1 $a = 4, \; 2x^2, \; -12x^3$

2 0.464 cm^2

3 $\tfrac{4}{3}, \; 3x + 4y - 18 = 0, \; x - 2y + 9 = 0,$
63.4°

4 0.45

5 a) $\theta = 2n\pi \pm \dfrac{\pi}{3} + \dfrac{\pi}{6}$

b) ii) $\dfrac{\pi}{4\sqrt{3}}$ cubic units

6
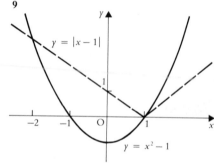

7

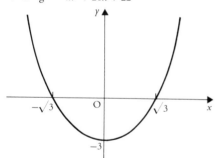

Positive m

a) $m < -\frac{4}{9}$ and $m > 0$

$m \in \left(-\infty, -\frac{4}{9}\right) \cup (0, \infty)$

equal roots if $m = -\frac{4}{9}$

$4x + 9y - 12 = 0$

8 a) $\left(\dfrac{1}{\sqrt{e}}, -\dfrac{1}{2e}\right)$, minimum

9 $3\pi\left(\ln 2 - \frac{1}{2}\right)$

10 1.434, 0.717

Paper

13

1 $y = \frac{1}{3}(8 - x)^{\frac{1}{3}}$

$\dfrac{2}{3} - \dfrac{x}{36} - \dfrac{x^2}{864}$

3 a) $(-1, \infty)$ b) $(-\infty, -8) \cup (-1, \infty)$

4 $f \circ g = 4x^2 + 20x + 22$

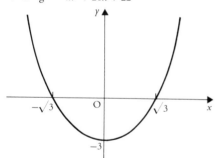

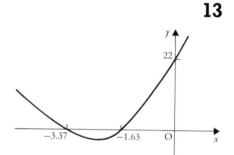

range of $f \circ g$ is $[-3, 166]$

5 a) $z = 2\sqrt{3} + 2i$, $w = -1 - \sqrt{3}i$

b)

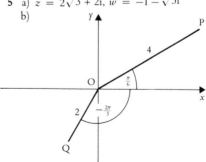

$PQ^2 = 20 + 8\sqrt{3}$

c) $\dfrac{z}{w}$: mod 2, arg $\frac{5\pi}{3}$

w^2: mod 4, arg $\frac{2\pi}{3}$

6 $\tan 67\frac{1}{2}° = \sqrt{3 + \sqrt{8}}$

7 max. of 16 at $x = 0$, min. of -16 at $x = 4$

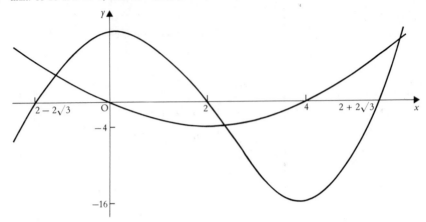

$2(1 - \sqrt{3}) < x < 2$ and $x > 2(\sqrt{3} + 1)$ $k > 16$

8 a) L_1 $\mathbf{r}_1 = 3\mathbf{i} + 6\mathbf{j} + \mathbf{k} + t(2\mathbf{i} + 3\mathbf{j} - \mathbf{k})$
     L_2 $\mathbf{r}_2 = 3\mathbf{i} - \mathbf{j} + 4\mathbf{k} + s(\mathbf{i} - 2\mathbf{j} + \mathbf{k})$
   L_1 and L_2 intersect at the point $(1, 3, 2)$
  b) $56.9°$
  c) $a' = -5$, $b = 6$, $C \left(5, \frac{13}{3}, 2\right)$,

$6.912\left(\sqrt{\dfrac{430}{9}}\right)$,

$\dfrac{15}{\sqrt{430}}\mathbf{i} + \dfrac{13}{\sqrt{430}}\mathbf{j} + \dfrac{6}{\sqrt{430}}\mathbf{k}$

9 $A = 2$, $B = 3$, $C = -1$,
   $2x + 5y - 17 = 0$

10 $2 - \sqrt{3}$

Paper   14

1 $13 \cos(\theta + 22.6°)$
   $g(x) = 12 \cos 2x - 5 \sin 2x + 15$
  a) max. value 28, min. value 2
  b) $78.7°$, $258.7°$
  c) $48.7°$ to $108.7°$

2 a) $\theta = \dfrac{12}{x} - 2$ c) 9 sq. units

3 $\left(-1, -\dfrac{1}{e}\right)$, $\left[-\dfrac{1}{e}, \infty\right)$

  b)

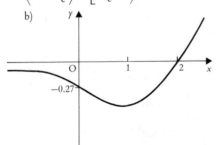

4 a) $\mathbf{r} = \mathbf{i} + \frac{1}{3}\mathbf{j} + t(-\mathbf{i} + \frac{2}{3}\mathbf{j} - 2\mathbf{k})$
  b) $(3, -1, 4)$

5 $y = 0$, $y + 2x = 24$, $y - 2x = -24$

7 117 minutes $[200(2 - \sqrt{2})]$

8 $y = -\dfrac{1}{x - 2} + \dfrac{4}{x + 6}$

$\dfrac{dy}{dx} = \dfrac{1}{(x - 2)^2} - \dfrac{4}{(x + 6)^2}$

$\dfrac{d^2y}{dx^2} = -\dfrac{1}{(x - 2)^3} + \dfrac{8}{(x + 6)^3}$

$\dfrac{dy}{dx} = 0$ when $x = -\frac{2}{3}$ or 10

max. of $\frac{1}{8}$ when $x = 10$

min. of $\frac{9}{8}$ when $x = -\frac{2}{3}$

9

$y = 2x^{\frac{1}{3}}$

$y = x^{\frac{1}{3}}$

$y = x^{\frac{1}{2}}$

$\dfrac{1}{\sqrt{2}}$

10 $2 \ln \left| \dfrac{8 - x}{4 - x} \right|$

$t = 2 \ln \left| \dfrac{8 - x}{4 - x} \right| - 2 \ln 3$

1.01 p.m.

Paper 15

1 a) $p = -1, q = 7$ b) $-7\mathbf{i}$

3 a) $x = \dfrac{1}{\ln a - 2}$ b) $x = \frac{1}{2}\sqrt{ae}$

4 max. Y is ≈ 2.27

5 a) $\mathbf{p} = (1 - \lambda)\mathbf{i} + \lambda\mathbf{j}, \mathbf{q} = (1 - \lambda)\mathbf{j} + \lambda\mathbf{k}$
$\mathbf{PQ} = (\lambda - 1)\mathbf{i} + (1 - 2\lambda)\mathbf{j} + \lambda\mathbf{k}$

$|\mathbf{PQ}|_{\min} = \dfrac{1}{\sqrt{2}}$ $|\mathbf{PQ}|_{\max} = \sqrt{2}$

c) $\frac{1}{7}\mathbf{i} + \frac{2}{7}\mathbf{j} + \frac{4}{7}\mathbf{k}$

6 a) $60°$ b) $45°$ c) $54.7°$

7 $\dfrac{dy}{dx} = \dfrac{2(x^4 - 1)}{x(x^4 + 1)}$,   min. at $(-1, \ln 2)$
min. at $(1, \ln 2)$

8 $69°$ **9** $2 \cot 2t$

10 a) ii) $-\frac{1}{34}$ b) 1.701 accurate value 1.773